FORCED TO PLOUGH

Bonded Labour in Nepal's Agricultural Economy

A report for
Informal Sector Service Centre
and Anti-Slavery International

ADAM ROBERTSON
AND SHISHAM MISHRA

ANTI-SLAVERY
INTERNATIONAL

इन्सेक
insec

Forced to Plough:
Bonded Labour in Nepal's Agricultural Economy

by Adam Robertson and Shisham Mishra

No. 10 in ASI's Human Rights Series

Published by Anti-Slavery International (London)
and Informal Sector Service Centre (Kathmandu)
Editors: Hugo Plowden, Mike Dottridge, Margaret O'Grady

Photographs: Adam Robertson
Design: Suzanne Perkins/Grafica
Maps: Adam Robertson and Oxfam
Printed by: Indreni Offset Press, Kathmandu

British Library Cataloguing in Publication Data
A catalogue record of this book is available from the British Library
ISBN 0 900918 37 3

Anti-Slavery International gratefully acknowledges the support of
Royal Norwegian Ministry of Foreign Affairs, and of Swedish International Development Assistance (SIDA). INSEC gratefully acknowledges the support of British Executive Service Overseas (BESO).

Acknowledgments

The authors would like to thank the many hundreds of people who gave up what little free time they had to answer questions in the course of the research. Special thanks are due to Depak Deo Bhatt (Thagil Social Development Association) who provided invaluable assistance in data gathering and in field work and to Ro Cole who read and made extensive and helpful comments on an early draft of this text. In addition, thanks are due to the following individuals and organisations who generously gave their time and support to this project: Anil Mishra, Sustainable Livelihood Forum (SLF), Child Workers in Nepal (CWIN), Dr Shiva Sharma (APROSC), Dr Ganesh Gurung, Mr Kesab Badal MP, Suresh Chaudhary, Gauri Pradhan, Krishna Kaphley, Rami Ram Pokharel, Rocky Ram Pokharel, Mohan Sing Nayak, Mr Bharat Karki (APROSC), Backward Society for Education (BASE), Mr Rishi Raj Lumsali, Satinarayan Chaudhary (TWS), Peter Powell, Mohan Sing Rathor, John Burlison (BESO), Harald Skar, Dilli Chaudhary (BASE), Inge Sagild (BASE), Puranbar Chaudhary (BASE), Sophia Powell, Ramanada Prasad Sing, Parshuram Tamang (NEFEN), Bal Bahadur Dangaura, Bijaru Chaudhary, Raj Kumar Tharu, Firolal Chaudhary, Erik (BASE), Puran Ram Dyal, Sita Ram Thapa (SLF), Dr Hari Bansh Jha, Pavan Kumar Ojha, Shesh Kanta Sharma (HURON-Baglung), Hem Mehar Shrestra (Ministry of Land Reform), Surya Battarai (INSEC), Biro Ram Lohar, Tara Niraula, Bikram Chaudhary, Jan Hopkins, Motilal Sharma Bhushal, Prashu Narayan Chaudhary and Omar Sattaur.

Contents

Contents

Introduction

Behind its image as a beautiful country for foreign visitors, life in rural Nepal is hard. Women bent double with their loads toil up the mountain paths, carrying fodder for the animals or manure for the fields. Men, barefoot and weather-beaten, work long hours with ox and plough, using age-old farming methods.

In most of the country transportation of goods and people is still by foot. Smoky wood fires are used for cooking and many hours each day are spent collecting the fuel. Good quality drinking water still remains a problem in many areas and the sources are often a long trek away. At night, with no electricity, lighting is by kerosene lamps, or candles, and each morning brings another day in the struggle to eke out a survival from the land.

It is these daily pressures which lead an increasing number of adults and children to migrate to the cities in search of another way of life. For some young Nepali girls, working in the sweat shop conditions of Kathmandu's carpet factories becomes a preferable alternative to the relentless grind of life at home.[1] The majority of farms are large enough only to feed a family for a few months of the year and the male members of the family are consequently forced to migrate in the hungry months for labouring jobs and send what they can back home.

Families have to stick together to survive and the extended family system is a common insurance against disaster. There is often a high degree of individual self-sacrifice for the sake of the family and community, and a tradition of neighbourly generosity and sharing of responsibilities.

All Nepalis recognise that village life is hard; fewer will admit that it is harder, a lot harder, for some than it is for others. But there is no escaping the fact that the rural community is also a divided one. It is divided along lines of gender, wealth and caste. The different groups may be tightly knit and all play their part within the system, but often the bonds that tie them together are based at best on patronage and at worst on exploitation. At the extremes this exploitation degenerates into slavery.

Powerful families still control rural society in large areas of the lowland plains and make use of bonded and forced labour from the villagers who live on their estates. In the western hills, bonded labour is closely associated with the caste system which is still very strong in the area and forces certain groups into servile status. Both these structures have led to a division of the community into landlords and serfs, high-caste and low-caste, exploiters and exploited.

The exploited groups, powerless to initiate change, have known no other way of life and have little option but to accept the status quo. They have instead rationalised their position by adopting a fatalistic approach to life, the last refuge of the oppressed. The landlords and high-castes are locked into the same system and justify their privileged status as a right of birth. The result is that, like many rural societies, both groups look on change with suspicion and abdicate responsibility for their actions to the dictates of tradition.

Bonded Labour and Slavery in Nepal

Although slavery in the Kingdom of Nepal was officially abolished in 1926, it is still common today, taking the form of debt bondage, or bonded labour as it is generally known in South Asia. Debt bondage was banned by the United Nations in 1956. According to the United Nations Supplementary Convention on the Abolition of Slavery, the Slave Trade, and Institutions and Practices Similar to Slavery (1956), debt bondage is defined as:

> *..the status or condition arising from a pledge by a debtor of his personal services or of those of a person under his control as security for a debt, if the value of those services as reasonably assessed is not applied towards the liquidation of the debt or the length and nature of those services are not respectively limited and defined.*[2]

Bonded labour is a problem common throughout the South Asian region. Laws banning bonded labour now exist in both India and Pakistan, but although Nepal has signed the Supplementary Convention on the Abolition of Slavery, the Slave Trade, and Institutions and Practices Similar to Slavery (1956), it has not yet introduced domestic legislation of its own to ban the practice.

The Government of Nepal now acknowledges that a form of debt bondage exists in the country, but considers the problem limited to five districts in western Nepal affecting only a few thousand people. This report sets out the evidence that the problem of bonded labour exists much more widely throughout the rural areas of the country.

Although there has been no survey of the number of people affected by debt bondage, it is estimated that they run into hundreds of thousands: men and women who have mortgaged their working life in lieu of interest on a loan they will never be able to repay. It is ironic that in many cases the original debt may amount to no more than the average family in the industrialised world would spend on a restaurant meal.

Nepal - Country Profile

Nepal is a country of immense diversity. This is as true of its geographical make-up, as it is of its ethnic composition. It is arranged into three distinct ecological zones running from east to west. Travelling north from the border with India one passes through the fertile lowland plains of the *terai* region, through the hills and valleys of the Siwalik and Mahabharat ranges, and up to the Himalayas themselves, their 8,000-metre peaks forming a jagged edge to the northern border with Tibet. While the hill zone occupies the largest area, the most fertile land is concentrated in the narrow *terai* strip. The mountain zone is sparsely populated, whereas 85 per cent of the population lives in the hills and *terai*.

The 1991 census records a population of approximately 18.5 million people speaking some 20 different languages. At various stages in the country's past, attempts have been made to incorporate Nepal's many different ethnic groups into a formalised caste system based on the Hindu model. A Hindu system of social stratification now exists in most parts of the country and is especially strong in the rural areas. The high-caste *Brahmin* and *Chhetri* are established as the most powerful groups in Nepal, together with the Newar high-castes of the Kathmandu Valley.

Until the mid-18th century, Nepal was divided into many small principalities. Even after that time, when King Prithvinarayan Shah unified the country in 1768, feudal politics still governed the lives of the peasants. The Shah dynasty which he founded lost power in 1846 to an elite family, the Ranas, who called themselves hereditary prime ministers. The royal family however was kept on to maintain a figurehead and the two families, the Shah Kings and Rana Prime Ministers, became interlinked by marriage.

Throughout the century of Rana rule, Nepal resisted contact with the wider world and gained a reputation for cutting itself off from developments outside its borders. It continued to be governed by powerful landed families who owed homage and service to the King and Rana Prime Ministers. Nevertheless, Nepal was not immune from the massive political changes which had overtaken the Indian sub-Continent in the post-war years, and the influence of its southern neighbour, India, had a particular impact.

In 1951 King Tribhuvan and the fledgling Nepali Congress Party, with the support of Nehru's India, attempted to introduce democracy to the country, and foreign diplomatic missions were given access to the capital. King Tribhuvan died in 1955 and was succeeded by his son Mahendra, who introduced a democratic constitution, and held an election, won by the Nepali Congress Party. However, this brief democratic interlude ended abruptly in 1960, when King Mahendra imprisoned the Prime Minister and assumed control of the government.

In 1962, King Mahendra imposed the Panchayat system, under which political parties were banned and the country was administered through a system of councils. Village councils elected members to district councils which, in turn, elected the members of the national governing body, the National Panchayat. Real power, however, rested with the King who appointed the Prime Minister, the cabinet and over 40 per cent of the members in the national chamber. Behind the scenes, the elite ruling families maintained their power through a system of patronage. This system was mirrored through the workings of the Panchayat councils all the way down to the village level where local powerful families continued to dominate rural life.

The Panchayat system was maintained by King Birendra, the present monarch. Although the direct power of the Crown was reduced after 1980, political parties continued to be prohibited and their leaders imprisoned.

In 1989, the underground opposition parties formed a coalition to press for a multi-party democracy with the King as its constitutional head. In February 1990 the Government attempted to put down non-violent protests with tear gas, mass arrests, torture, beatings and brutal force. In the months of unrest which followed, many hundreds of people were killed when police opened fire on crowds of unarmed demonstrators. Eventually, the 'people's movement' faced down the strong-arm tactics of the Government and ushered in a new era of democracy.

The two main opposition parties which had led the people's movement, the Nepali Congress (NC) and the Communist Party of Nepal, United Marxist and Leninist (CPN UML), formed an interim government. The new government remained faithful to the two main demands which had been at the heart of the people's movement, the desire for democracy and for human rights. They acceded swiftly to the UN International Covenant on Economic, Social and Cultural Rights, and the UN International Covenant on Civil and Political Rights, and undertook a programme of democratic reform culminating in the drafting of a new constitution. Multi- party elections were held in 1991.

The Nepali Congress won the 1991 election and stayed in power for the next three years. However, in July 1994 the Government lost a key vote in Parliament, prompting the Prime Minister to resign and precipitating another general election. This resulted in no clear majority, but the party with the most seats, the Communist Party of Nepal, United Marxist and Leninist (CPN UML) was asked to form a government.After the CPN UML had been in office for nine months, the Nepali Congress, having forged an alliance with the third largest party, the Rastryia Prajatantra Party (RPP)and the Nepal Sadvawana Party, (NSP), brought down the Communist government in September 1995. The coalition partners maintained an uneasy relationship characterised by infighting and internal power struggles until early 1997 when a new coalition government, this time comprising members of the CPN UML, RPP and the NSP, took the reins of power.

Nepal: Districts referred to in report

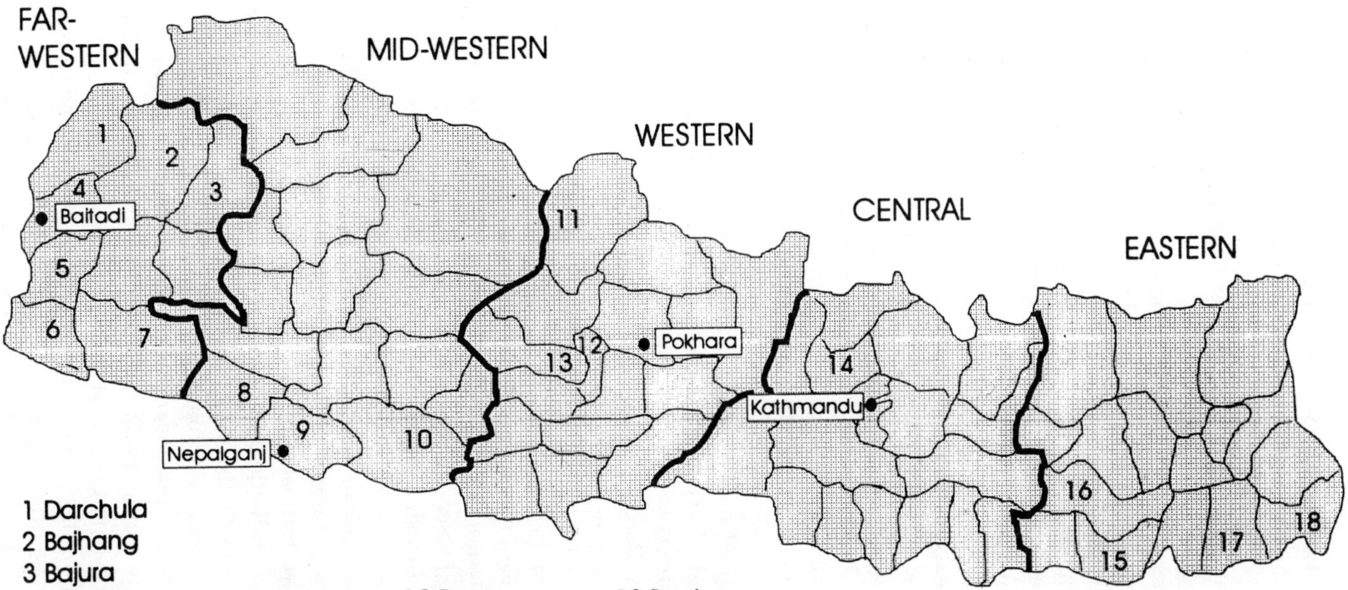

1 Darchula
2 Bajhang
3 Bajura
4 Baitadi
5 Dadeldhura
6 Kanchanpur
7 Kailali
8 Bardiya
9 Banke
10 Dang
11 Mustang
12 Parbat
13 Baglung
14 Sindhupalchok
15 Saptari
16 Udayapur
17 Morang
18 Jhapa

The Fight Against Slavery in Nepal

The *Kamaiya* system of bonded labour in the western *terai* was first noted by anthropologists in the 1960s,[3] but it is mainly since the democratic reforms of 1991 that action has been taken to expose the system. This has been led by a handful of activists from non-governmental organisations (NGOs) working in the area. They have launched successful programmes to empower the affected groups and to press for wider government action.

One of the NGOs at the forefront of this work is the Informal Sector Service Centre (INSEC), an NGO dedicated to the protection and promotion of human rights and democracy in Nepal. In 1991, INSEC carried out the first survey of the *Kamaiya* system and on the basis of this research it started literacy and human rights classes in *kamaiya* villages to help organise the bonded labourers and help them assert their rights.

INSEC operates through a network of grass-roots organisations working in 32 of Nepal's 75 districts. Through this network INSEC received information indicating that in addition to the *Kamaiya* system there were similar exploitative labour practices operating in other areas of the country.

In 1994 the London-based human rights organisation Anti-Slavery International (ASI) and INSEC carried out further research in five different districts to try to find out more and to identify the groups affected by slavery. Based on the initial findings of this research, both organisations decided to work together to focus the attention of the Government of Nepal and international aid agencies on the much wider problems of bonded labour in Nepal.

A small team was set up to carry out the additional research required. This consisted of one researcher from INSEC and one researcher from ASI. The ASI researcher was sponsored to work with INSEC by a volunteer agency, British Executive Service Overseas (BESO).

The research was restricted to four geographical areas of the country where bonded labour was believed to be occurring: the hills of the Far-Western development region, the hills of the Western development region, the western *terai* and the eastern *terai*. Information was gathered by group and individual interview with a total of over 300 agricultural labourers from 11 different districts over a six-month period from September 1995 to February 1996. In addition, local landlords, other experts and NGO and government officers were interviewed. The report also draws upon other published and unpublished research carried out by INSEC over the past four years.

The focus of this report is necessarily on the bonded labour contract, which is a male preserve, and the situation of women in bonded labour families has not been studied in detail. Nonetheless, from information

Nepal: Major ecological zones

Map reproduced with the kind permission of Oxfam.

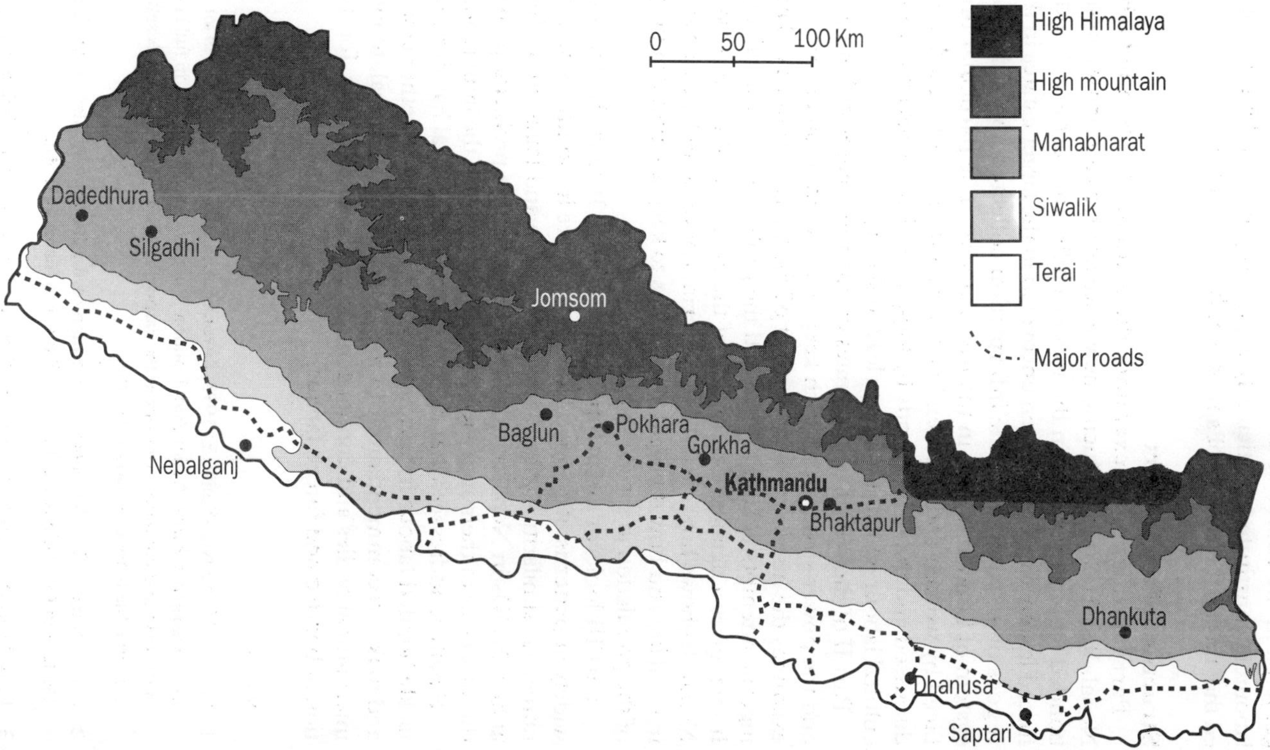

gathered in the course of the research, it is obvious that women bonded labourers constitute a particularly exploited group, and it is recommended that this be the subject of special study in the future.

Structure of this Report

Part I of this report examines the evidence for bonded labour in Nepal. It starts by introducing new evidence for caste-based systems of bonded labour prevalent in the western hills, and forms of bonded and forced labour which have been uncovered in the larger estates in the *terai*. It goes on to summarise the information collected about the *Kamaiya* system of bonded labour in the western *terai*. The particular problems faced by women and children in all these systems of bonded labour are discussed and the section ends by looking at daily-wage systems of agricultural labour which sometimes lead to slavery.

Part II of the report concentrates on the root causes of bonded labour and the power structures which keep the system in place. First of all, it examines the historical development of the current pattern of landholding, the attempts at land reform and the problems of rural indebtedness. It then gives a detailed description of the evolution of the caste system in Nepal. It shows how caste-based discrimination is still a common factor in rural life and how the rigid system acts to keep a significant proportion of the population in servile status.

Part III looks at the prospects for change. It starts by looking more widely at patterns of bonded labour elsewhere in South Asia, noting the numerous similarities with cases in both India and Pakistan. It then looks at how NGOs in Nepal have begun to address the problem, pointing out the successes of their approaches, as well as some of the limitations.

Finally it addresses what the Government of Nepal has done so far to tackle bonded labour in agriculture. Part IV draws together conclusions and makes recommendations to the Government of Nepal, to interested internatonal bodies and donor agencies, and to local NGOs, for the abolition of bonded and forced labour.

NOTES

1 Johnson, V and Ivan-Smith, E, 1995. *Listening to Smaller Voices: Children in an Environment of Change*, Action Aid, London. Research indicated that, in full knowledge of the risks and conditions of work in the carpet factories, some young girls saw the work as preferable to their situation at home.

2 United Nations Supplementary Convention on the Abolition of Slavery, the Slave Trade, and Institutions and Practices Similar to Slavery (1956).

3 Bista, D. B, 1967. *People of Nepal*, p119, Ratna Pustak Bhandar, Kathmandu.

Part 1
Bonded Labour in Nepal

During the ploughing season bonded labourers work up to 20 hours a day, working through the night if necessary.

CHAPTER 1

The *Haliya* System: Bonded Labour in the Western Hills

The word '*haliya*' comes from the word '*halo*' meaning plough in Nepali. Strictly speaking '*haliya*' therefore means 'one who ploughs' but it is understood to have the broader sense of an agricultural labourer who works on land other than his own.

There are many different systems of agricultural labour in Nepal that are all loosely described by the word '*haliya*'. Some are simply a system of wage labour (See Chapter 5), others amount to slavery under international definitions. There are different regional variations on the word and someone called a *haliya* in the hills might be called a *hali* or *haruwa* in the terai (plains) with no real difference in meaning. In some Far-Western districts *haliya* is interchangeable with the word *kamaiya* to describe debt bonded workers from the Tharu indigenous group.[1]

The Effects of Caste and Feudal Structures

The act of ploughing the land is seen as being exclusively a low-caste occupation. In the strict application of caste laws, a *Brahmin* or *Chhetri* who ploughs may risk losing his high-caste status. Nowadays, the punishment is never actually enforced but, nonetheless, high-caste people generally avoid ploughing. This is particularly true in the Far- and Mid-Western hills where the caste system is still strong, but even in other areas of Nepal ploughing is seen to be indicative of low social status. Those high-caste people who do plough, do so only out of economic necessity and many do not admit to it openly, as it is still a cause of shame. The same situation is also found among the more 'Brahaminised' ethnic groups who, in an attempt to raise their own status, adopt the customs of the high castes.

It is this caste factor which makes ploughing different to all other types of agricultural work and explains why it is treated separately by the farmers. The powerful custom demands that only untouchable castes or low status ethnic groups plough the land as *haliya*.

The term '*haliya*' is associated most closely with a system of debt bondage in caste-based communities, which is prevalent in the western hills. Small or medium sized high-caste farmers are not supposed to demean themselves by engaging in ploughing, but often cannot afford to pay market rates for agricultural labour. The dilemma is resolved by use of debt bondage to secure unpaid labour from low-caste labourers. This is particularly important at peak times in the agricultural calendar when otherwise wage labour would be most expensive. This arrangement is a product of a rigid caste system and the low-caste *haliya* is bound to work not only by debt but by custom. It violates the United Nations Supplementary Convention on the Abolition of Slavery, the Slave Trade and Institutions and Practices Similar to Slavery (1956) to which Nepal is a signatory.

Information collected by Informal Sector Service Centre (INSEC) indicates that such systems are found in the following districts of Nepal: Baitadi, Bajhang, Bajura, Dadeldhura, Darchula, Dhawalagiri, Gorakha, Gulimi, Mustang, and Parbat. Other evidence suggests that the problem is widespread throughout the hilly regions of Western, Mid-Western and Far-Western Nepal. [2]

The landlords are generally from high-caste *Brahmin* or *Chhetri* groups or from the higher status ethnic groups such as *Gurung* and *Magar*. The *haliya* are usually of 'untouchable' castes, such as *Kami* (and *Lohar*), *Dhamai* and *Sarki* castes. Very rarely high-caste people may also work as *haliya*. (See Chapter 7 for a full description of the Nepali caste system).

Debt Bondage

The UN Supplementary Convention on the Abolition of Slavery, the Slave Trade, and Institutions and Practices Similar to Slavery (1956) defines debt bondage as:

> *...the status or condition arising from a pledge by a debtor of his personal services or of those of a person under his control as security for a debt, if the value of those services as reasonably assessed is not applied towards the liquidation of the debt or the length and nature of those services are not respectively limited and defined.*

The work that *haliya* do is not applied towards the liquidation of the debt.

In Baitadi District where Jhari Ram Lohar (cited below) lives, the local going rate for agricultural labour is NR 48 (US$0.85) per day. If Jhari Ram were paid at this rate, in one year he would be able to pay back the principal and interest on his loan (at 5 per cent per month), pay for his own food (currently supplied by the landlord) and having made these payments he would still have NR 6,600 (US$119.00) left to support his family. Instead, he has been labouring for six years and can see no end to

Jhari Ram Lohar, who agreed to talk to researchers on condition that his identity was disguised.

Jhari Ram Lohar

Jhari Ram Lohar has been working as a *haliya* for six years for a debt of NR 3,000 (US$53.00). He lives with his family in Koleuda, a tiny hamlet of 'untouchable' houses in Melauli Village Development Committee (VDC), Baitadi District, Far-Western region. He has a wife, two sons and three daughters and has his own house and a small kitchen garden of less than 0.1 hectare. The landlord he works for as a *haliya* is high-caste, has 2.5 hectares of farmland and also works in government service.

Jhari Ram has been juggling debts all his adult life but eventually could not avoid taking work as a *haliya*. He now sees no way out of the system.

I borrowed NR 3,000 (US$54.00) *for survival, for food. It's a big problem to feed the family. I work just to pay the interest on the loan. When I need more money I go back to the landlord. The loan is increasing all the time. God knows how I will ever pay it off. If I don't work, then I have to pay five per cent interest a month.*

The hardest months are Jestha and Ashar (May, June and July). *Sometimes I have to work from dawn until it's very dark. If necessary I start work at midnight and continue through. I do all the work on the farm, ploughing, harvesting and the rest. I am busy for about eight months of the year, but I have to be around to do other jobs too like portering, taking goods to the market (a day's walk to the nearest road).*

Usually I get two meals a day and at harvest time I get a sack of grain (40-50 kilogrammes, worth NR 200 or US$4.00).

I can't go for wage labour or the landlord would kick up a fuss and say 'You owe me money so you must work for me'. Usually the landlord is good. He treats me with respect. It is the system that is very unfair. We are under pressure from all sides because of the loan and because of our caste. [3]

his life of servitude. His debt has not changed.To argue that *haliya* in this system are not bonded through debt is untenable. There is a legitimate argument which says that because of poverty, land-hunger and disguised unemployment in the hills, *haliya* are at least lucky to get two meals a day. Those who employ this argument should not deceive themselves; in effect they are arguing that these human beings are lucky to be slaves.

A recent (1994) Nepal Rastra Bank survey of rural credit identifies one source of rural credit coming from landlords:

> *The role of the landlord as a money lender is interlinked with input and output markets. Since the landlord's basic interest is in cultivation of his land, his lending is not directed toward earning interest. His interest rather lies in the regular assurance of the labour supply. On account of this his loan is interest-free. But the interest of such loan is indirectly earned through reduced labour price.*

Among the types of borrowings from the landlord the most prevalent is the bonded labour type.[4]

Although this statement was corroborated by ASI and INSEC's research, it only tells half the story. To call the loan 'interest free' is misleading. If the *haliya* fails to turn up for work when he is needed, he is charged punitive interest at anything up to six per cent per month. As such, the loan is given simply to bind the *haliya* to the landlord. Furthermore the 'labour price' is commonly reduced to the provision of just one or two meals a day.

The *haliya* sometimes works for a debt of just a few thousand rupees. In effect, he is forgoing the local daily wage for his labour to pay interest on this debt. Based on a survey of 149 *haliya* in three VDCs of Baitadi District (carried out by INSEC and ASI) it was possible to calculate that by accepting to work for food alone and forgoing the local daily wage, in lieu of interest payments on the debt, the *haliya* were in effect paying interest at rates exceeding 10 per cent per month [5]

The work that *haliya* do is not limited by time, and their duties are not defined. Debts are passed on from father to son and with them the responsibility for working as *haliya*. Even though the landlord may change, some families have been bonded by the same debt for up to four generations. Dhami Ram Lohar, a *haliya* from Radana Basadi VDC in Baitadi District, has no illusions on this count:

> *There's no way out of the system. You have to keep on working right up to your last breath, that's the only way out. The only other way out is to force your own sons to be* haliya *in your place.*
>
> *I hope that my children will not work as* haliya. *I will work very hard to*

see that they are educated. But every year I say 'this year I will get out' and every year I end up staying.[6]

The work continues from boyhood all through adult life until the *haliya* is no longer capable of hard labour and so passes on the responsibility to his sons. According to ASI and INSEC's survey, the amount of the loan varies from NR 200 (US$3.60) to NR 50,000 (US$900.00) and the loan is made for an indefinite period. In fact, far from being expected, repayment is actively discouraged. *Haliya* have to pay the loan back in full in one lump sum or the landlords don't accept it. Because their earnings are so meagre, it is usually quite impossible to pay off the loan in this way. The only way a *haliya* can usually escape is by selling whatever land and possessions he has (if they are not already mortgaged) thus making his family homeless. Ram Mani, 22, of Salyantar VDC, Parbat District, took this desperate course of action:

The loan was increasing and all the time there was such pressure on me. I sold my land (0.02 hectare). *I have a wife and small child and now we are squatting on government land living in a hut. Some of the high-caste people are trying to chase me off.*[7]

Rather than paying off the loan, more often a *haliya* will require further credit from the landlord just to buy food or clothing. If, however, no further credit is available, a *haliya* can try to change his landlord. This is done by getting a bigger loan from another landlord and then paying off the old one. The longer a *haliya* has been working, the bigger the debt tends to be.

In the same way that *haliya's* work is not limited by time, their duties are also not defined. When there is little agricultural work left to do on the landlord's farm, the landlord may 'lend' the *haliya* to work for his relatives, or keep the *haliya* available for portering jobs, carrying produce to or from the local market towns. Contracts indicating the length of time a *haliya* has to work, his hours or duties, are unheard of and debts are rarely formally recorded.

Seasonal Work and 'Semi-Bonded Labour'

In the western hills the landlords do not usually possess large farms. Although there are still some large landholders, the average farm size in the area is 0.8 hectare.[8]

During the off-season when there is very little agricultural work, the *haliya* are not needed and are not given any daily food. They must search for whatever other work they can find in order to survive, and many of the men migrate to India or the larger cities. However, they are bound, by debt and custom, to return when they are required by the landlord. The seasonal nature of the work has led some to call this type of *Haliya* system 'semi-bonded labour'. This phrase is unsatisfactory and has no basis in international legal definitions. The *haliya* always remains under the control of the landlord and he is bonded through debt. It is simply that, rather than feed the haliya when there is no work for him to do, the landlord finds it more profitable to cast him off to live on his own resources for this period; he does this in the sure knowledge that the *haliya* is bound to return when required. There is also an assumption that being allowed to migrate, the *haliya* are able to bring home riches from the big cities in India. But although the earnings from migrant labour can be important, the net contribution to the household budget is marginal in many cases, as Dhami Ram Lohar, a *haliya* from Radana Basadi VDC, Baitadi District, explains:

> *There are four brothers in our family. We all go to work in India for four months of the year and together we bring back about 6,000 Indian Rupees* (US$170.00). *But it costs to travel and live there too and the rest of the family* (11 other members) *have to take on a loan to survive while we are away. By the time we pay back the loan and interest and everything else there's nothing left. Why do we go? Well, when there's no work here at least there's one less mouth to feed.*[9]

For some, *haliya* migration is made impractical because they have to stay working for their employer for up to 10 months of the year.

Landlord-*Haliya* Relationship

Usually, relations between *haliya* and the landlords are no better or worse than those between any high-caste and low-caste groups. Some landlords use degrading language to their *haliya* and sometimes violence is threatened. On rare occasions these threats are carried out. Khajendra Ram Lohar of Selena VDC has been working for eight years as a *haliya* for a debt of NR 3,000 (US$54.00). He is now 32:

> *It was during the time of rice transplanting. I was working for many days in mud and water right through into the evening. My legs and feet became so painful and swollen because of the water. I couldn't go near the water any more. Then one evening the landlord ordered me to go and wash the oxen in the stream. I showed him my legs and said I couldn't go near the water again because I was in pain. I refused to go so he beat me. Some other people stepped in and stopped him in the end. I didn't wash the oxen but the next morning I went back to the fields again and nothing more was said.*[10]

If *haliya* become ill they are required to send a replacement on the days they are sick. This will usually be another male family member. If no one is available they are charged interest on the days they miss and do not receive their daily ration of food. Some *haliya* interviewed in the course of the research were allowed time off in the evenings to go to a health post if they were injured.

Variations on the *Haliya* System

Each district has a slightly different system. In some villages *haliya* may get two meals a day, in another they may get none at all. Sometimes old clothes are supplied to the *haliya* and sometimes they may be given a sack of grain at the end of the harvest. In the survey carried out by INSEC and ASI 9.4 per cent of the sample had no real debt to the landlord but had bonded themselves to the landlord simply in order to get the daily food ration. Apart from the varying rates of remuneration, more fundamental differences can be found. In some areas *haliya* were bonded for the ploughing season only, that is for just four months of the year. In other villages *haliya* are given a loan on condition that they are available for work for the whole year. They work for the four-month ploughing season for which they receive only a daily meal. After that they are paid at the local daily wage rate for the days they work. From these wages the *haliya* are expected to pay interest on the loan at five per cent per month.

Where farm sizes are particularly small, several landlords may club together and share a *haliya* between them. Large land owners however may elect to pay their *haliya* by giving them a small patch of land. Even though some of these variations may not start out as bonded labour, it is clear that in most cases *haliya* were not able to pay off the debts and sooner or later the systems gave way to the classical forms of debt bondage outlined above.

It is clear that the different forms of *haliya* system in the western hills are really just variations around a common system of debt bondage based on caste. The following chapter goes on to look at the situation in the plains where *haliya* are enslaved in systems based on serfdom.

NOTES

1 see Chapter 6.

2 INSEC, 1994, *Human Rights Year Book 1993*, INSEC, Kathmandu, Chapter 8.

3 From research carried out by ASI/INSEC in January 1996.

4 Nepal Rastra Bank, 1994. *Nepal Rural Credit Review, Final Report*, p 84, Kathmandu.

5 A study of minimum wages for agricultural workers in Nepal was conducted by the Ministry of Labour, the Centre for Economic and Technical Studies and the

International Labour Organisation in Autumn 1995.

6 From research carried out by ASI/INSEC in December 1995.

7 *ibid.*

8 Central Bureau of Statistics, 1994. *National Sample Census of Agriculture Nepal 1991/92 - Analysis of Results,* His Majesty's Government National Planning Commission Secretariat, Kathmandu.

9 From research carried out by ASI/INSEC in October 1995.

10 *ibid.*

CHAPTER 2

Debt Bondage in the Plains

Close to the Indian border in Saptari District, Eastern Nepal, is the landlords' village of Kailadi. It is set amongst thousands of hectares of prime farmland watered by the Sapta Kosi River. The village itself is hidden behind leafy bamboo glades; television aerials and satellite dishes can just be seen between the palms. A tour around the village reveals expansive concrete and brick villas built around a series of lily ponds and small lakes. These are the homes of about 100 of some of the biggest and richest landlords in Nepal. Some are reported to control holdings of over 1,400 hectares in this and other districts.

The village of Mushari, although only five minutes walk across the fields, is a complete contrast. It is a settlement of *Musahar* people built on land owned by one of the landlords in Kailadi: a collection of 36 very simple huts made from a frame of interlaced sticks and twigs, where a total of 300 individuals reside. Ram Dev Sadar is 54 and has been living here since his grandfather's time.

> *He (*the landlord*) gives us land for our houses so that we can work for him. Our job is digging ditches. Then we carry the mud from one place to another to make the fields (*building dykes and terraces for rice cultivation*). It takes a whole day for one man to move 100 mon (four tonnes) of mud for say 20 metres. That's our main job, but we also do all sorts of other agricultural work, ploughing and harvesting and the rest. That keeps us busy for six months of the year. We get some breakfast when we work and we get a daily wage paid in rice* (worth NR 10-12 or US$0.18-0.22).

The children in the village run naked and barefoot. Despite the rich agricultural farmland which surrounds them, they are malnourished and underfed. None go to school, many work watching cattle and in return receive some rice for a meal. *Musahar* are considered to be the lowest of the low-castes in the area and have 'untouchable' status (see Chapter 7). They are also known by the caste name of *Khataves*. Traditionally this caste is 'employed in ditch-digging and other agricultural work.[1] It is impossible for them to find work apart from the tasks prescribed by their caste. They are not allowed near the high-caste landlords' houses.

> *A raise? They wouldn't give us a raise. If we asked for a raise they wouldn't give us work ever again. They'd tell us we had to move. It is very hard to manage; there are many big families in our village with mouths to feed. Even if we went for a loan we wouldn't get one. We survive by getting work outside. The younger men have to leave to work on the sugar cane plantations in Punjab (*India*).*[2]

Serfdom

Serfdom sounds like an old-fashioned concept connected with the Middle Ages in Europe, or 19th century Russia. However, it is a modern reality in Nepal. The UN Supplementary Convention on the Abolition of Slavery, the Slave Trade, and Institutions and Practices Similar to Slavery (1956) defines serfdom as:

> *...the condition or status of a tenant who is by law, custom, or agreement bound to live and labour on land belonging to another person and to render some determinate service to such other person whether for reward or not and is not free to change his status.*

In the plains large landholders require large amounts of cheap and dependable agricultural labour in order to maximise their profits. Farm sizes are generally larger than in the hills and landowners may have one or more low-caste settlements on their estate. Sometimes these settlements may amount to no more than four or five households, but on the bigger estates whole villages may effectively be owned by the landlord. The low-caste families are forced to provide labour to the landlords at a fraction of the market rate; otherwise they risk eviction. It is this which forces them to work for the landlord. Work obligations may extend to all family members including women and children and for certain traditional roles the villagers may be required to give their services with no payment at all.

The *haliya* working in this system are serfs. Their role as *haliya* is dictated by their caste, a status which they are unable to change. Their caste demands that they perform agricultural labour and other forms of employment are not open to them. They are dependent on the landlord for their daily food and shelter. Although the younger men of the village can migrate when there is little agricultural work to do, they leave their dependents behind and are obliged by custom to return when they are needed. Breaking this custom would mean that the landlord would evict the family from his property making them homeless and destitute. Thus, it is a practical impossibility to leave his service, and they are forced to accept whatever conditions are imposed.

Accommodation

Because the land on which the village is built is owned by the landlord, the villagers are constantly in fear of eviction. The land given to *haliya* to build houses is usually very small. They live in rudimentary shelters with patchy thatched roofs and wickerwork walls daubed with mud plaster which they build and repair themselves. Even though they build and repair their dwellings themselves, they do not own them. Because the landlord has rights over the land, he assumes rights over whatever is built on that land. This even extends to produce grown by the *haliya* on unused patches of ground next to their huts. A *haliya* from Saptari District, Eastern region told of how, over several years, he had spent what little free time he had nurturing an orange tree which he had planted on the plot next to his shack. Once the tree matured and started to bear fruit, however, the landlord stepped in and claimed the fruit and tree as his own.[3]

Another Form of Bondage

Often *haliya* in the plains will also take on loans from their landlords and become bonded like their counterparts in the hills. More frequently a different form of debt bondage prevails, where workers are paid only at the end of the year. This forces them to take on loans from the landlord to pay for food. Whether or not they are obliged to pay interest on these loans depends on the landlord. Without exception they are not allowed to leave the landlords' service until the debts are repaid. When pay day finally arrives at the end of the year, they find that most of their salary disappears to pay off the debt and the next yearly cycle begins. Having trapped the *haliya* in this system the landlords can force down the wage rates. The *haliya* interviewed in the course of this research said that they knew of no other system and accepted this regime as their fate.

S.S. (who wishes to remain anonymous) is a landlord in Toxila VDC, Udayapur, and has several *haliya* families on his land. He calls some of the *haliya* in the village together to talk with the visiting researchers. He explains:

> *They do the agricultural work and stay all the year round, but they are free to leave if they wish. In return they get three meals a day and 12 sacks of rice* (approximately NR 2,000 or US$36.00) *in kind after the harvest. Then they get a monthly salary of NR 700* (US$13.00). *They get sick leave and medical facilities up to about NR 1,000* (US$180.00) – *but of course that depends on the landlord. Some landlords round here don't give sick leave and take out days missed from the wage.*

Kumar B. Daji is 20 years old and lives as a *haliya* on land belonging to S.S.'s neighbour.

> *I started working when I was four. My father was a tailor and he died. Then my mother left and we had to work to survive. I work the whole year round. I start early in the morning (6 am) and finish in the evening (6 pm), but we have two hours off in the afternoon [when it becomes too hot to work]. I get 12 sacks of rice* (equivalent to NR 2,000 or US$36.00) *each year and food three times a day and NR 200* (US$3.60) *per month. No, it's not enough to survive on. The rest of the family go for wage labour. They get paid more.* Kumar's wages are roughly one-fifth that of an ordinary agricultural labourer in the area. Worse still, he is paid only at the end of the year. S.S. confirms that this is the system. *The wage is paid annually. But if they need money the landlord gives an interest free advance; it's not really a loan. It is also true that some of the inexperienced* haliya *get a smaller wage... No, that's right, they are not free to leave until they pay back the advance.*[4]

Domestic Slavery

Some *haliya* are given no land for a house at all, and instead sleep in the stable or under the eaves of the landlord's house. Typically they are required to perform domestic work as well as work in the fields and are known as *haliya nokar* (servants). In this situation the landlord controls virtually every aspect of their lives, including where they sleep and what they eat. As domestic servants they are on-call 24 hours a day. In spite of this increased work load they receive very low remuneration. As the example of Ram Dev Paswan shows, *haliya nokar* represent one of the most exploited groups encountered in the course of the research.

There are about 100 landlords' houses in the high-caste village of Kailadi and each will employ at least one person like Ram Dev. Some will employ several, many of them young children of seven or eight. Living on the landlord's land, debt bonded, 'untouchable' and employed in domestic service as well as agricultural work, such *haliya* are under pressure from many different sides. The landlord's control over them is complete.

The phenomenon of *haliya nokar* is common in many parts of the country. INSEC has reported practices where 'the rich keep poor labourers at home to work as tillers or herdsmen more or less as slaves' occurring in 20 out of the 75 districts in Nepal.[5]

In this system the earnings of the *haliya* are so low that they will not be able to feed their dependents. Very often this results in the whole family helping in the work in order to get a meal. Children will watch the cattle and women may be asked to carry out a whole range of domestic work and other chores on demand.

Ram Dev Paswan

Ram Dev Paswan doesn't know his exact age but he is in his early twenties. He comes from a small village of low caste and landless people called *Chanti Gau*, in Kailadi VDC, Saptari District. He is recently married but does not yet have children.

I have been a haliya nokar *(servant) for 10 years. My father is one too. We work for the same landlord, Raja Babu, who owns this village. My father is at the house now – feeding the cattle.*

Raja Babu is the respectful term for the landlord Ramananda Prassad Sing who owns a large estate of some 200 hectares in this and other districts. He has a large house in Kailadi village.

I do agricultural work and I do domestic work too. I come back to the village to eat my food but the rest of the time I stay at the big house. I get up at 4 am every morning and work until about midday. Then I come back here to the village to eat; it only takes 15 minutes to walk. I have an hour or so for lunch, depending on what needs to be done. Then I go back to the house or carry out the other duties in the fields. I can be working up to 8 pm looking after the irrigation pumps in the evening. Then I get another meal.

*I get 50 kilogrammes of rice (*un-husked*) a month (*worth NR 200 or US$3.60) *and I get two sets of old clothes each year. I don't get any other food. We sell some of the rice to buy dal (*lentils*), and we glean some other vegetables from the fields. It's not enough to survive on.*

My father took on a loan originally. That is why we are working like this. I had to take a loan as well. I have borrowed two Mon (80 kilogrammes) *of rice* (worth NR320 or US$6.00) *and NR 4,000* [US$72.00] *as cash. I have to pay 20 kilogrammes of rice in interest each year and NR400* (US$7.00) *interest on the loan. The interest is taken out of my wages but usually I can't pay it and I have to borrow more and the interest is added to the debt. I have to keep working because of the debt. This is no life living under someone else. If I get sick I have to pay for the days I miss but people with proper jobs don't have to do that, do they? They get sick leave. How will I ever pay this debt off? I'll always have it. My children will have it too.*[6]

NOTES

1 Bista, D.B, 1967.p113, *op. cit.*

2 From research carried out by ASI/INSEC in January 1996.

3 *ibid.*

4 *ibid.*

5 INSEC,1994.*op. cit.* Chapter 8.

6 From research carried out by ASI/INSEC in January 1996.

CHAPTER 3

The *Kamaiya* System

> *We work as* kamaiya *and* bukrahi *for a debt. We don't have our own land and we survive by working for the landlord. If I miss one day's work the debt increases. I have a sick child, what should I do? Go to work, or stay and look after the baby? I don't even have time to clean his soiled clothes. We are human beings and we should have time off when we are sick.*
> Lumali Tharuni, woman bonded labourer (*bukrahi*), Bardiya District, Mid-Western Nepal.[1]

The word *kamaiya* is in many senses synonymous with the word *haliya* and shares many similarities with the patterns of exploitation described in previous chapters. However, it is necessary to consider the *Kamaiya* system separately for three important reasons. Firstly, it has sprung from quite different origins, coming originally from the indigenous culture of the Tharu people of the western *terai*, and consequently has its own terminology and variations.

Secondly, of all the systems of bonded labour in Nepal it is the most publicised, the best researched, and the only one which is being directly addressed by government and NGOs. There is consequently much better information available. Thirdly, it is the only system which allows for the virtual buying and selling of bonded labourers by their landlords in a form of veiled slave trade.

The Tharu

The *Kamaiya* system was first noted in the 1960s, by anthropologists studying the Tharu people in the western *terai*.[2]

Most of the research done so far has concentrated on five districts in western Nepal: Kanchanpur, Kailali, Bardiya, Banke and Dang. The *Kamaiya* system is, however, known to exist also in Surkhet, Nawalparsi, Rupendehi, and Kapilvastu Districts. The system is characterised by bonded labour and serfdom and is thought to affect over 100,000 individuals. Ninety-five per cent of those affected are Tharu.

With a population of 1,194,224 (6.5 per cent of Nepal's population) the Tharu are one of the country's largest ethnic groups. They are indigenous to the *terai* region where the vast majority are still located, across 22 districts from east to west. They are particularly numerous in the western *terai*. Seventy-nine per cent of Tharu above the age of six (when children should start school) have never attended school.[3]

Historically, the Tharu were feared by outsiders and the land they inhabited was infested with malaria to which the Tharu were said to have acquired some natural resistance. Consequently, they were left to develop in comparative isolation for many centuries. It is only in the last 100 years that they have come into direct contact with neighbouring communities. For the Tharu this contact has brought disastrous results.

Originally the Tharu's traditional lands extended well into northern India but in the 19th century the forests they lived in were rapidly eroded from the south to supply the British Raj with timber for railway sleepers. In the same period in Nepal, however, the Shah rulers and later the Ranas faced a dilemma. On the one hand they saw the malarial jungle as a useful southern barrier between themselves and their expansionist neighbour, while on the other they needed to extend their tax base. For these reasons direct colonisation was limited to quasi-military settlements, while the local Tharu headmen were empowered to collect taxes for the Crown.[4]

The system of indirect taxation, however, gradually changed and the tax collectors were increasingly drawn from high-caste Nepali-speaking people from the hills who had connections at court. These tax collectors were called *jamindars*.

In addition, large tracts of land in the *terai* were given away by the King as rewards to favoured courtiers or generals to ensure their loyalty to the Crown. The land granted in this way was called *birta* land and the recipients called *birtawalas* (see Chapter 6). They had power to collect revenue from the Tharu people who worked the land and in return they paid tribute to the Crown.

The new administrators would come only occasionally to oversee their property and collect revenue. The longest period they would stay would be for two or three months in the dry season so as to minimise the risks of malaria.

The introduction of *jamindars* and the granting of *birta* lands saw an expansion of the farmed area of the *terai* and a depletion of the Tharu's traditional forest lands. Ironically, it was the growing number of landless Tharu who, because of their resistance to malaria, were used as labour to clear this land and make wider cultivation possible. The Tharu thus became the instruments of their own undoing. By the beginning of the 20th century the pattern had been established by which the Tharu people would ultimately become disenfranchised, but it was not until the introduction of a malaria eradication programme in the 1960s that the area was opened up to mass migration from the hills.

Powerful landed families from the hills were able to increase their holdings in the newly opened forests and further marginalise the indigenous Tharu. Although a small proportion of Tharu became landlords themselves, and adopted the ways of the majority high-caste society, the vast majority of the Tharu were left in an increasingly vulnerable position and ripe for exploitation.

The Evolution of the *Kamaiya* System

There is some evidence to suggest that the *Kamaiya* system developed initially from a survival strategy used by the Tharu to help insure against the risks of sedentary farming.[5] Whatever its origins, the new landlords quickly converted the system into a highly exploitative one where large landholders keep *kamaiya* to work their land in the same way that they keep oxen, or other farm livestock.

TARA PRASAD THARU is a landlord in Nambasta VDC, and lives with his extended family in a large brick-built complex, which is the central point in the village. Around it are the barns for the cows and oxen and the clusters of *kamaiya* shacks. Within the house itself *bukrahi* (female *kamaiya*) are carrying water and washing plates from the previous night's meal. Tara Prasad has 40 *bigha* of land (27.2 hectares) which is worked by seven pairs of *kamaiya* and *bukrahi*.

We keep kamaiya *to plough the land. I don't know if it's right or wrong or what, we have land and we need* kamaiya. *It's as simple as that. There isn't another system that works. If I give them money they will just spend it on alcohol. They don't know how to deal with money.*

They do the harvesting work and all the field work. At the moment they are very busy working in the fields, but it's not true that they work at night – they start work at 4.00 am. I only employ couples; if it's just a kamaiya *they get a reduced share. Life for* kamaiya *has improved. Before, there were less people and more land so the* kamaiya *had to work 24 hours a day, more or less. Now they don't have to work so hard.*[6]

A government commission in 1995 identified some 17,000 *kamaiya* families in the five districts of Kanchanpur, Kailali, Bardiya, Banke, and Dang.[7] Of these about 50 per cent were working in debt bondage, prohibited by the United Nations Supplementary Convention on the Abolition of Slavery, the Slave Trade, and Institutions and Practices Similar to Slavery.

Further, 58 per cent of *kamaiya* families were found to own no land of their own for a house and were therefore dependent upon the landlord for a roof over their heads. The main elements of the *Kamaiya* system are similar to those of the other forms of bondage described in previous chapters. *Kamaiya* are usually paid in kind, and usually bonded by debt to their employer. The main difference however is that they are obliged to stay on the land all year round. A verbal contract to this effect is traditionally made with the landlord during the festival of *Maghi* in mid-January each year.

Kamaiya are usually given a payment in rice after the harvest; this payment is known as *bigha*. In addition they are given a portion of other grains, salt and oil, again handed over in one lump sum to supplement the rice. The total amount is gauged to be just enough for survival, and sometimes slightly less.

Control over the *kamaiya* is maintained by keeping them in a perpetual cycle of debt. The debt increases to pay for medicines, additional food, and any other necessities. Often the landlords inflate the debt still further and charge the *kamaiya* for any day's work which they miss through illness. Likewise, any damage to a piece of equipment or domestic animal for which the *kamaiya* is held responsible is added to the debt, often without the *kamaiya's* knowledge. A *kamaiya* who tries to pay off his debt can expect to be presented with a long list of expenses and spiralling interest so that, in fact, the money is extremely difficult to pay off. In this way the owner of the debt becomes the owner of a human being, the *kamaiya*.

The fundamental difference between the *Haliya* system and the *Kamaiya* system is that the latter allows for the buying and selling of one person by another. The debt attached to each *kamaiya* not only bonds him to his landlord but, in effect, gives him a cash value in Nepali rupees. The *kamaiya* thus represents a tangible asset which the landlord can sell to others. These transactions are often carried out between landlords during the traditional time of *Magh* (January). Although ostensibly it is the debt which is being bought and sold, it carries with it the power of ownership over the *kamaiya*. As such, the debt becomes merely a veil for what in fact is a form of slave trade. Sometimes brokers act as middlemen between the *kamaiya* and the landowner. According to one landlord in Kanchanpur:

> *There are many middlemen, most of them are Tharu. They sit with them, give them alcohol and pretend to be their friends, but when they are deep enough in debt the brokers sell the debts over to the landlords and that's how they become* kamaiya.[8]

MANI RAM CHAUDHARY is 25 years old, married with one daughter. He was originally sold by his cousin after his father's death.

I have been working here for two years now but I started working as a kamaiya *11 years ago. Originally I came from Dangadi, Kailali District. My father owned land in Dangadi, but I was young when he died and my cousin transferred all the land into his name and he threw me out.*

My cousin sold me as a kamaiya *for NR 500* (US$9.00). *The loan starts from there. I got food and old clothes and two quintals (200 kilogrammes) of grain a year. It was very bad, hard work. I was never free like before. Then I moved on to another landlord. He was a lawyer. My cousin came and took a loan from him in my name without telling me. I only found out when my cousin died that I had a debt of NR 7,000* (US$126.00). *Now the debt is NR 14,000* (US$252.00).[9]

One System or Many?

There are many variations on the *Kamaiya* system, displaying a whole spectrum of exploitation. There are differences according to whether the labourer is in debt or not, and whether the *kamaiya* owns his own house or patch of land on which it is built. There is also considerable variation from district to district and landlord to landlord in the amount and type of remuneration paid.

The most exploitative cases occur when the *kamaiya* is both in debt and without land of his own. In these circumstances he is obliged to live on the landlord's property. According to the latest government research, some 24,000 individual family members from five districts in western Nepal are in such a situation.[10]

When a *kamaiya* has no land of his own, his landlord usually allocates a small plot next to the farmhouse on which to build a shack. On large properties, in addition to the barns and stables which surround the landlord's house, there is invariably a collection of *kamaiya* shacks. The *kamaiya* are totally dependent on the landlord for food, credit, and accommodation. If the *kamaiya* wish to leave their landlord, they have to abandon their homes in addition to paying back debts.

In most cases the debt owed by a *kamaiya* is relatively small, below NR 2,000 (US$36.00), but with no cash income, this is virtually impossible to repay. The debt is inherited and passed down from father to son.

Landlords use verbal threats and humiliation to intimidate the *kamaiya*. Although *kamaiya* are reluctant to report physical abuse, it is clear that violence is also used. In Kanchanpur District, out of those *kamaiya* who responded to INSEC's questions regarding ill treatment by their employer, 14 per cent said that they had been subjected to beatings. [12]

VIOLENCE AGAINST *KAMAIYA*

Kushi Ram Tharu lives and works as a *kamaiya* in Rajapur VDC, in Bardiya District. He is a young man of 24 but has been working for the same landlord, Raja Ram Sing, since his childhood. His family has been bonded since his grandfather's time. His grandfather originally took on a loan of just a few thousand rupees. He was initially reluctant to relate his story for fear of reprisals:

I still remember the day I was beaten totally by Raja Ram. My father was working as a* kamaiya *in his house. When I was asked to work I refused, but I was forced to work. I said "I am very young, I cannot do such hard work", but he forced me and said I had no choice. I ran away but was caught on the border and taken back. He beat me until my body became swollen, I could not move from the bed for four days. When I was taken back he threatened me. He said he would kill me and throw me in the river if I tried to run away. [13]

Similarly Sunurua Tharu, a middle-aged man from the same VDC, related how he too had been subjected to violence and humiliated by his landlord.

Once I was carrying the rice crop to Rajapur by bullock cart and the cart went out of control. For that he beat me in front of everyone. I could not speak, I had no voice. I just shed silent tears.

Another time I was beaten for not making a rope bed. I arranged all the ropes but was not given any wood to make the frame and legs with. How could I make the bed without the wood. [14]

Generally, landlords require that the *kamaiya* offers his wife or other female relative for domestic work and other agricultural duties. The woman, whether she is a daughter-in-law, wife or mother, is referred to as a *bukrahi*.

The *kamaiya* do all the ploughing and heavy field work. They usually work for a 12-hour continuous period receiving their meals in the field. During harvest and ploughing periods, however, they often continue through the night working 20-hour days for weeks at a time. In the off-season they are either given other work or are loaned or hired out to their landlord's friends or relatives. They are not free to work for wage labour elsewhere.

The *bukrahi* work in the landlord's home as domestic servants, but also process farm produce and work in the fields at peak times. They work the same hours as the men but in addition have to carry out a second round of domestic duties within the home when they return.

NOTES

1 Mishra, S, 1994. *Voice of the Voiceless* (video), INSEC, Kathmandu.

2 Bista, D.B.1967.p.119, *op cit.*

3 Extrapolated from Central Bureau of Statistics, 1992. *National Population Census of 1991*, His Majesty's Government National Planning Commission Secretariat, Kathmandu.

4 Skar, H.O, 1991. *Nepal Indigenous Issues and Civil Rights: The Plight of the Rana Tharu*, pp 15-16, Norwegian Institute of International Affairs, Oslo.

5 Rankin, K, in H.O. Skar (Ed.), *Tharu and Tarain Neighbours*, in press.

6 From research carried out by ASI/INSEC in October 1995.

7 Landless Settlers' Problem Resolution Commission, 1995. *Survey on Landless Bonded Labour* (Kamaiya)*: A Report*, p2, His Majesty's Government of Nepal, Kathmandu.

8 From research carried out by ASI/INSEC in October 1995.

9 *ibid.*

10 Landless Settlers' Problem Resolution Commission, 1995, p12, *op. cit.*

11 INSEC, 1992. p 68, *op.cit.*

12 Mishra, S, 1994. *Women Bonded Labour in Nepal*, Unpublished Research, INSEC, Kathmandu.

13 *ibid.*

CHAPTER 4

Women and Children in the Bonded Labour System

Regardless of class or caste, women constitute one of the most oppressed groups in rural Nepal. In comparison to men, women work longer hours. The average work burden per day for women is estimated at 10.8 hours in comparison to 7.5 hours for men.[1]

Most of the labour-intensive agricultural work is carried out by women – sowing, transplanting young rice plants, weeding, harvesting and processing the grain. The only exception is ploughing which is exclusively men's work. Women are paid a lower daily wage rate for their labour than men, and they receive a smaller portion of food, even though their calorific requirements may be higher because of the longer hours worked.

In rural areas girls are not considered a priority for education and start work at a younger age than boys, helping their mothers with the burdens of child care and other domestic duties. Their lack of access to education is just one of many reasons why women are very rarely represented in the local decision-making process.

Women do not have formal rights to own or inherit land and so have restricted access to credit. Marriage increases a woman's dependency (although on her husband rather than her father or brother); divorce is legal and can be requested by a woman on several grounds, but she does not have any rights over the former family home or the property she shared with her husband.

Women in the *Haliya* System

Ambika Ropa Sarki looks much older than her 35 years. She is low-caste, a woman, and widowed which is a disastrous combination in Nepal:

> *I have no land, no house, and no husband. I have four children to look after and one has a broken arm. What can I do?*

The only option for her is to work as a labourer. Her 13 year-old daughter looks after the other children, the youngest of whom is five years old.

Ambika works in Trupeshaur VDC, Udayapur District. She does any work she can find. As an 'untouchable' it is difficult to get domestic work and so she mainly works as an agricultural labourer. As a woman she gets paid half the men's wage even though she takes on all kinds of work in the fields except ploughing. This amounts to NR 15 (US$0.26) per day and three meals. As a widow in Nepal she is stigmatised and unlikely ever to find another husband; without a husband it is hard to find a house. Landlords need people to plough; women do not plough, so they do not get a place to live.

At the moment someone who has gone to Kathmandu has lent me their hut to live in. I don't know where I will go when they come back.Because I don't have any property I can't get a loan but when I am desperate I can sometimes ask for an advance on my wages.[2]

Women are not called *haliya* because they very rarely plough, but nonetheless, they are heavily involved in agricultural work. All the family helps during the rice crop. The wives have to work so that the *haliya* get finished more quickly and can work on their own patch of land or kitchen garden. They don't get paid.[3] According to Dill Bahadur Dhading this is the system which operates in Parbat District:

The women have to work. It's traditional during the harvest. They get a meal when they work. They also work wage labour at other times but they only get half pay because they don't plough.[4]

The women bring in the harvest, carry manure to the fields, and gather fodder for the animals. They work on the landlord's land as well as their own, if they have any. This, of course, is in addition to all the domestic chores that have to be done at home and the many hours spent collecting firewood, preparing meals and looking after young children. When in the off-season their husbands migrate in search of work, all the burden falls upon the women in the family who are expected to make ends meet as best as possible.

In Hardiya (eastern *terai*), when the men migrate the women usually stay at home, picking up what work they can and looking after the children. The money sent home by their husbands does not always reach them, however, and they are often forced to take on a loan until their husbands' return.

Buchri Das Chaudhary says that when her husband goes to Delhi or Punjab he tries to send money home by a courier or friend.

He usually sends a few hundred Indian rupees back but often the whole *amount doesn't get here. Last year he sent 1,000 rupees (*US$28.00*) but I only ever got 200 (*US$5.70*) because the courier takes it.*

It is no exaggeration to say that wives of *haliya* work all their waking hours. But it is extremely hard to imagine just what that means. Buchri Das Chaudhary tries to explain:

Rest time? We have no rest time, just time to sleep. We do all the work except ploughing, we do all the domestic work too in the landlords' houses, carrying water, cooking, sending their children to school. We do all the cooking, washing, field work; the winnowing, threshing and so on for our own homes as well. Then we have the children to take care of.[5]

For six to twelve days after the birth of a child women are allowed to rest, but it depends on the time of year and what work needs to be done. Sometimes they have to go out and work in order to bring in the harvest just three days after giving birth.

Women in the *Kamaiya* System

Working for bangles – *a bukrahi* from Dang District, Mid-Western Nepal

Within the *Kamaiya* system women are double victims of servitude. *Bukrahi* (women bonded labourers in the *kamaiya* system) are not parties to the contracts struck by their male partners and the landlord. They do not receive direct remuneration for their work but receive only the daily meal. Sometimes they may be given old clothes or a length of cloth by their landlord or his wife.

Young unmarried girls aged between eight and 16 are sometimes taken to work exclusively in their landlord's house and are referred to as *orgoniya* in the *Kamaiya* system. Such girls are vulnerable to sexual abuse. The sexual abuse of women bonded labourers is widely acknowledged but individual cases are seldom reported. In one infamous case from a village in Bardiya District, every newly-wedded *kamaiya* bride had first to sleep with the village landlord on her wedding night. The landlord concerned was later humiliated when the story came to the attention of the wider public, but he has never received any formal punishment.[6]

If a *bukrahi* is widowed her future becomes very uncertain. For example, 32 year-old Rampati Chaudhary has five small children. Her husband died at work in the landlord's fields. Both of them had been working for the landlord, C K Sing, for four years. Now he is demanding that she leaves the *bukhra* and pays back the debt. He explained:

> *"It is not beneficial to keep a single* bukrahi *without a* kamaiya *because she is not able to plough the land."*

RAM KUMARI THARU is 19. She is *bukrahi* to her landlord, Ayotesa, who has 18 *bigha* of land (12 hectares) in Nambasta VDC, Banke District, Mid-Western Nepal. Altogether the landlord has five *kamaiya* families on his land. Ram Kumari is paired with her father, Lohari Tharu. She took over as *bukrahi* from her mother at the age of 13.

The family debt is NR 2,000 (US$36.00) and they do not know if the debt is increasing or decreasing.

I work cleaning in the landlord's house and now that it is harvest time I am all day in the fields until dark. I just have a half-hour break for lunch. I should get paid for my work – I work just as hard as the* kamaiya*; I feel very sad about it.

Ram Kumari does not get any direct remuneration. Neither does she get any clothes. Each year she receives a dozen glass bangles for her labour.[7]

Rampati is completely destitute and does not know what to do. She says her only option is to beg. [8]

This pressure to find a female partner for a *kamaiya* leads to a practice of child marriage. Young sons will marry olde_ women so that their wives can work as *bukrahi* and give their mothers some relief. Eighteen year-old Asrani works as a *bukrahi* in the house of Raja Ram Chandas in Bardiya District. She was married with 13 year-old Santa Bir so that she could replace her mother-in-law as a *bukrahi*. Asrani works as *bukrahi* while her husband goes to school where he is studying in fifth grade.[9]

Children in the Bonded Labour System

Tara Ram Lohar is 13 years old and has started to help out his brother Madan Ram Lohar as a *haliya* in the landlord's fields.

> *I started working a year ago. I still go to school but it's getting more and more difficult. I work digging the corners when my brother is ploughing and I help out at harvest time. When I work I get a meal.*

It is rare for the children of *haliya* to go to school and very few complete a primary school education. Less than half of *haliya* families send any of their children to school. [10] The pressures on the family are such that children must work for their food when possible or help out with the domestic duties. The amount of work given to girl children is particularly heavy.

Even though there is no specific contract between the family members and the *haliya's* landlord, in many cases the whole family is considered to be bonded to the landlord and thus may be called upon to provide labour at any time.

Generally the children of *kamaiya* are employed as cowherds to watch the landlord's cattle, for which they receive nothing but a meal. Sometimes, however, they become bonded in their own right. This happens particularly after the death of one or both parents when the child inherits the burden of a debt. Some six to seven per cent of *kamaiya* contacted in INSEC's 1991 survey started working as *kamaiya* when they were below the age of 10; a further 20-25 per cent started working from between the ages of 10-14.[11] Orphaned *kamaiya* children become extremely vulnerable to the worst sort of exploitation. The Tharu NGO, Backward Society Education (BASE) identified 5,920 children aged 5-12 years old who were *kamaiya* in their own right in the early 1990s, about 400 of whom were orphans.[12]

Twelve year-old Ashram Chaudhary is a *kamaiya* in Shreepur Majhgoan. His father died, his elder brother ran away and the responsibility for his mother (aged 55) and two younger brothers (aged 10 and 8) rests with him and his adult sister (aged 20). He does not even know about the NR 14,000 (US$252.00) debt he has taken on. He is obliged to work for food alone, without receiving any sort of payment. His mother and sister work in the landlord's home while Ashram grazes cows with his brothers. He is subjected to beatings by his landlord and says that his sister is being sexually exploited. [13]

NOTES

1 Oxfam, 1990. *Nepal Programme Review*, Oxford.

2 *ibid.*

3 From research carried out by ASI/INSEC in October 1995.

4 From research carried out by ASI/INSEC in December 1995.

5 From research carried out by ASI/INSEC in January 1996.

6 Mishra, S., *op.cit.*

7 From research carried out by ASI/INSEC in October1995.

8 Mishra, S., *op.cit.*

9 *ibid.*

10 Information from ASI/INSEC survey.

11 INSEC, 1992. *Bonded Labour in Nepal under* Kamaiya *System*, p 54, INSEC, Kathmandu.

12 Backward Society Education (BASE), 1994. *Tharu Education for Transformation*, Third Year Report (submitted to DANIDA), Dang. Appendix 10/10/95.

13 INSEC, 1992, p 106, *op.cit.*

CHAPTER 5

Daily-Wage *Haliya*

BIR LAL RAI

Bir Lal Rai is 62 years old. His face is drawn and lined with age. He wears a pair of tattered shorts, a battered corduroy cap and is muffled in a blanket to protect himself from the night air. For the last five years he has been living in Tupeshaur VDC in Udayapur District. His shack is built on squatters' land near the banks of the Sapta Koshi river which drains the eastern Himalayas. He has a wife and four children.

***I was born in Bojpur, then I went to Patrai before ending up here. I have no land of my own and so I work as a* haliya.**

***I do all the agricultural work and any other labouring jobs if I can get them. For labour work I get NR 30* (US$0.50) *per day and a meal. For ploughing the rate is NR 25* (US$0.40) *for half a day because it is such hard work. For planting and all the rest they give me 12* mana *of rice* (NR 30 or US$0.50).**

***Yes I have a debt, let me see... it's about NR 1,200* (US$21.00). *At this time of year there is no work in the fields so I have to take on a loan just to survive. Life is hard, that's just how it is. When one moneylender dries up I try to search out another. I have loans from three different people at five per cent per month. At the moment I can just make the payments.*[1]**

Where labour supply is plentiful and cheap, landlords employ wage labourers by the day when they are needed. This practice is found throughout Nepal but is particularly common in districts where there are large numbers of impoverished landless settlers. *Haliya* working for a daily wage are, in many respects, more free than those who are employed under the other systems since there is no direct mechanism to bond the *haliya* to the landlord. Instead, the landlord can rely on the abject poverty of the landless or virtually-landless labourer to ensure a steady supply of labour. Ironically, it is precisely because there is a relative surplus of labourers that the landlord does not need to use debt bondage to meet his needs for cheap labour.

Where there are enough other employment opportunities in the off-season, the *haliya* can survive. Even if *haliya* are forced to take on debts from money-lenders to tide them over, these can be managed, and the interest payments met with money earned from migrant labour in India or Kathmandu. Often, however, unforeseen circumstances, or lack of alternative employment opportunities, force *haliya* to pledge their labour against a debt and the daily-wage system starts to give way to the more pernicious forms of bondage. It is this aspect of the daily-wage system, the fact that it is often the precursor to bonded labour, that makes it relevant for discussion here.

Wage Labour in the *Terai*

The landless in the *terai* come from many different castes and backgrounds. Some are from 'untouchable' castes who have migrated from the hills; some are from low-status tribal groups resident in the *terai*. Some are even high-caste farmers who have lost their land. Poverty being a great leveller, it is even possible to find *Brahmins* working as *haliya* amongst the landless of the eastern *terai*.

If they cannot get a piece of land from a relative on which to build a house, most will squat illegally on the forest fringes or on low quality land prone to flooding, by the banks of the major rivers. This is a precarious existence not only because of the environmental conditions but because the illegality of their tenure leaves them open to pressure from the local authorities.

Hardiya Village, Udayapur, Eastern *Terai*

There is a great variation in the prevailing rates for wage labour from village to village. In Hardiya VDC, Udayapur District, about 20 kilometres east of Tupeshuar where Bir Lal Rai lives, most of the daily-wage *haliya* are Tharu. Some have a small amount of registered land for a house and are thus more secure than their neighbours in Tupeshuar. The rates are consequently higher. The prevailing rate is NR 30 (US$0.50) per day but, in addition, the *haliya* receives three meals (equivalent to an extra NR 10-15 or US$0.25). Most of the *haliya* in Hardiya work there for a maximum of six months of the year. Sometimes landlords make short-term contracts for three to six months, in which case the *haliya* receives a lower wage to offset the increased security of employment. Only one *haliya* was working under a contract lasting 10 months: he was to be paid 720 kilos of rice a year (equivalent to NR 3,600 or US$65.00) and a daily meal. During the off season most of the *haliya* of Hardiya migrate to India or Kathmandu in search of labouring and portering jobs.

In Kathmandu the men can earn NR 60 (US$1.10) a day, and in Delhi or Punjab they get the equivalent of NR 90 (US$1.60). Always assuming they can find the work, they are usually able to break even. In some cases daily-wage labouring *haliya* families from Hardiya have been able to get loans to start animal husbandry with some success.

Wage Labour in the Hills

In the hills the majority of daily-wage *haliya* come from low-caste groups, in particular *Lohar*, *Kami* (blacksmith), *Sarki* (leather workers), and *Damai* (tailors). In contrast to people from the *terai*, *haliya* very often have some land of their own (up to approximately 0.1 hectare). In theory, this gives them a little more security, even though the land is often mortgaged. The other major difference in the western hills is that here caste discrimination is very strictly applied.

Silmi, Parbat District, Western Region

Silmi is a *Sarki* (leather worker) village in the hills, an hour's walk from the Kusma-Pokhara road in Parbat District. There are currently 42 families living there. No one works in the traditional area of leather work any longer. An increasing number of the younger men migrate to India to work as domestic servants or watchmen, sometimes staying away for a period of one or two years. Others work as porters, some on the road and some work as seasonal *haliya*.

Most of the families have a small piece of land (approximately 0.1 hectare) which produces enough to last the family for one or two months of the year. However, there is a high level of debt ranging from NR 5,000 to NR 35,000 (US$90.00–635.00) secured on the land. Very often the *haliya* will continue to work their mortgaged land as share croppers, giving half the crops they produce to the money-lender in lieu of part of the interest on the loan.

The caste system is rigidly enforced and caste discrimination practised. Custom demands that the lower castes carry out ploughing work for the whole community. Chardra Nepali, 55, has been working as a daily-wage *haliya* since he was 12 years old. For a 9–10 hour day he gets NR 30 (US$0.50) and one meal:

> *We are discriminated against because of our caste, and we have to work as* haliya. *None of the high-caste landlords plough. They just drink tea and don't work at all.*

In the hills, even if the *haliya* are not bonded by debt to a landlord, they are generally heavily in debt to the local money-lenders. As a last resort they will finally turn to their landlord for a loan. They then start to work to pay the interest on that debt and become drawn inexorably into work as bonded labourers as described in the previous chapters. In the *terai* the majority of daily-wage *haliya* own no land at all and their search for a roof often brings them to live and labour on a landlord's land. Thus, the system has many feudal characteristics.

NOTES

1 From research carried out by ASI/INSEC in January 1996.

Part II
The Causes of Bonded Labour

CHAPTER 6

Landlords and Landlessness

To have any chance of changing the entrenched social structures which have led to bonded labour in Nepal it is crucial to understand how they have evolved. This chapter describes the historic development of the current system of land holding and the failure of top-down strategies of land reform to address the massive problems of land hunger and rural poverty. The following chapter looks at the evolution of the caste system and its effect in providing a social justification for exploitation.

The Feudal Legacy

At present, the state retains control over the land and receives tax from the individuals who farm it. The present pattern of landholding, however, is still very much influenced by the feudal structures which were in place right up until the 1950s. These structures gave considerable powers to the local land-owning elite and despite attempts at land reform, these powers have endured at the village level.

In 1952, government controlled land accounted for only about 50 per cent of land holding and in addition there were three other feudal forms of ownership: *birta* land, *jagir* land and *rakam* land.[1]

Birta Land

Birta comes from the Nepali word *bir* meaning hero or warrior. The Shah kings and then later the Ranas granted huge areas of land to favoured generals and powerful local leaders not just to reward them for their valour in the battlefield but to secure their loyalty to the Crown. This land was called *birta* land and all the cultivators paid revenue directly to the feudal lord or *birtawala* who in turn paid tribute to the King. In 1952 this system accounted for about 700,000 hectares of land or 36 per cent of the total cultivable area of Nepal.

Jagir Land and *Rakam* Land

The word *jagir* in Nepali means official employment. During the period up to 1951 very often employees of the Crown would be assigned land by the King as payment for their services. They would therefore receive payment directly from the tenant farmers who lived and worked their fields. Similarly, *rakam* land was given to employees of the state as an alternative to cash, the only real difference being that recipients of *rakam* land carried out different duties which were usually of a manual nature. Together *jagir* and *rakam* land accounted for a further 150,000 hectares or eight per cent of the total area.[2]

Jamindars

The revenue from *birta*, *jagir*, *rakam* and government-owned land was collected either by the landlord himself or, more usually on the big estates, by intermediaries. These revenue collectors were called *jamindars*, and were generally empowered to deal with all aspects of land administration as well as rent collection. Even if they were acting as intermediaries for an overlord, their powers were extensive and they were often wealthy land owners in their own right. Nowadays, the title *jamindar* (or *zamindar*) is applied to any large land owner in Nepal as is the case in many parts of northern India, Pakistan and Bangladesh.

The feudal system demanded that all such landlords spend much of their time at court, dealing with the affairs of state where the King could keep a close eye on them. They would consequently visit their estates only periodically to collect rents. They had extensive powers over the peasants who cultivated their lands and were able to set whatever levels of rent suited them. The villagers living on their land had the status of serfs and the landlords could demand unpaid labour and other services from them.

After the collapse of the Rana regime in 1951, and in the spirit of a new egalitarianism, land reform was high on the new government's agenda. They introduced a series of far reaching legislative measures which abolished in turn *jagir*, *rakam* and *birta* systems of land ownership. Later the reform was continued under the non-party Panchyat system with the introduction of the 1964 Land Reform Programme.

The Failure of Land Reform

Ostensibly the Land Reform Programme had good intentions. It attempted to impose ceilings on the amount of land which individuals could own in order to redistribute land to the landless and victims of natural disasters. It also introduced a system of tenancy and land registration which aimed to remove the exploitative intermediaries (*jamindars*) and give greater security to those who were actively engaged in working the land.

Despite some early achievements however, a formal review of the programme in the 1980s concluded that:

The absence of dependable implementation machinery, lack of firmness and clarity in government policies, and considerable influence of the feudal elements severely constrained these measures.[3]

The basic mistake of the government of the time seems to have been to over-estimate the power of the state to claw back control from the local land owners. The former *jamindars* and *birtawalas* were well equipped to protect their land from the imposition of ceilings. They knew how to evade confiscation by registering land in the name of other family members or even servants, and if they themselves did not occupy powerful positions within the administration they frequently retained family contacts who did. The result was that only about one per cent of the original 600,000 hectares of land targeted for redistribution was actually redistributed.[4]

The landlords also resisted the introduction of new tenancy rights, seeing them as a further threat to their property. They were aided in this by the absence of land records and the fact that those charged with implementation of the scheme very often came from the same social grouping. It is estimated that nearly half of the potential number of rural workers eligible for tenancy rights were never even identified by the authorities and of the 1.8 million who were, only 300,000 eventually received certification.[5]

There is little doubt that many farmers eligible for the scheme were put off registering their tenancy rights with the authorities as a result of persuasion and direct intimidation by the landlords. As far as local people were concerned, the large landed families of the area remained in control and retained the traditional powers of patronage. The government in Kathmandu was much too distant to be seen as a real alternative to this local power base. Without the backing of the local elite the tenancy rights became little more than pieces of paper. There were, in addition, weak clauses in the legislation relating to tenants' rights so that even if tenants tested their rights in court, the landlords, with their knowledge of the legal process and their influence over the judiciary, were able to evict claimants from the land. Faced with these facts of life many tenants who did register their rights were later persuaded into 'voluntary surrenders'.

At the same time as the reforms were being partially implemented, the government started a major resettlement scheme in the *terai*. This strip of fertile land bordering the north Indian states of Uttar Pradesh and Bihar had been covered largely by forest and characterised by a high incidence of malaria. Although the land had been opened up from the 19th century onwards, the process was rapidly accelerated by malaria eradication campaigns sponsored by the World Health Organization in the 1950s and 1960s.

The intended beneficiaries of the resettlement scheme were supposed to be the rural poor, the landless and victims of natural disaster, but again the powerful landed classes were able to take advantage of the development. Consequently, families which already owned land in the hills were able to build up large estates in the newly accessible fertile soil of the *terai*.

The Panchayat Government of the time also had an ulterior motive in the resettlement scheme and planned to use it as a political device. At that time the *terai*, with its easy access to India, was being used as a sanctuary by political dissidents and opponents to the regime. In order to break up this potential hotbed of subversion much of the new land was awarded to Panchayat supporters and former loyal members of the armed forces. This was regardless of whether they met the programme criteria or not.

If the Land Reform Programme is to be judged by its ability to reach its own, admittedly ambitious targets, it can only be considered a failure. The unequal distribution of land remained largely unchanged and some landed families were even able to increase their share. According to official figures in 1991 five per cent of landlords controlled 26 per cent of the total agrarian land area.[6] However, the figures, show great variation from district to district. Some small-scale surveys have put the landlords' share as high as 40 per cent.[7] Certainly, landlords have developed very sophisticated systems of disguising the amount of land they hold from official investigation. Informally, some landlords admit to owning estates of 200 hectares or more, even though the legal maximum for a single owner is 18 hectares.

Marginalisation and Landlessness

A direct result of the concentration of land in the hands of a few is the marginalisation of the peasantry. Of course, the continuation of the feudal system depends on the availability of a cheap and compliant workforce. At present there is no shortage of labour, as an increasing number of families find their land insufficient to feed them through the year. Most farming is for subsistence and of the 2.74 million households owning some land, approximately 70 per cent own less than one hectare.[8]

At present levels of productivity a holding of less than one hectare is not sufficient to provide enough food for an average Nepali family.[9] Inheritance laws which divide land among all the male heirs on the death of their father result in an increasing fragmentation of landholdings and a decreasing average farm size. Although yields might be improved with the introduction of modern production inputs such as fertilisers and irrigation, these options are generally uneconomic on such small and fragmented holdings.

Growth in the agricultural sector has consequently been slow in comparison with other South Asian countries and is outstripped by population increase. To date, what growth there has been has come about mainly by increasing the amount of land being farmed and so is unlikely to be sustained. The net effect is land hunger with an increasing population dependent on a declining resource base.

Data on Bonded Labour

The National Sample Census on Agriculture for 1991/92 notes that 1.2 million holdings employed some form of agricultural wage labour for at least part of the year. In addition it notes that:

> *Work done in exchange for labour or otherwise without pay, provides an additional important component of farm labour but is not considered here.*[10]

Many of the workers who fall into this category are likely to be bonded labourers under the *Haliya* system or the *Kamaiya* system. However, the National Sample Census made no effort to estimate the number of labourers affected.

Landless agricultural labourers are extremely vulnerable to bondage and serfdom. The majority of landless agricultural labourers are already working as *haliya*. Data about the numbers of landless agricultural labourers is not collected by the population census. Rural households without any land are likewise excluded from Nepal's National Sample Census on Agriculture (NSCA). The only figures available come from a high-level government commission which carried out research into the problem of landlessness in 1995. It concluded that approximately one million households, though dependent on agriculture, owned less than 0.1 hectare of land and were effectively landless.[11] About half that number did not even have enough land to build a house of their own. Based on this figure it estimated that there were some two million landless agricultural labourers.

Many of these labourers have resorted to squatting illegally on unused government land by the major rivers or the forest fringes and work as daily-wage *haliya* (see Chapter 5). Others have offered themselves as *haliya* or *kamaiya* to local landlords in return for food and a patch of land to build a shelter. From there it is a rapid slide to bondage and serfdom. The only remaining factor in the equation is debt.

Indebtedness

In a survey of rural credit carried out by the Nepal Rastra Bank[12] in 1994, 50 per cent of households reported an outstanding debt. Most of this money was used to buy extra foodstuffs, pay for medical expenses, clothes and for religious festivals and marriages. Because loans are not available from banks and other institutions for these purposes, the vast majority of loans are taken from landlords and other traditional money-lenders. There is a disproportionately higher dependence on informal credit amongst the poorer households. The same survey found that among farmers with less than one hectare of land, loans from money-lenders accounted for four times the amount which they borrowed from banks.

Some households may acquire loans from friends and relatives at low rates of interest. For the poorest families without rich connections, however, such soft loans are hard to come by and generally interest rates are extremely high, ranging from two to six per cent a month.[13]

The high rates of interest, coupled with the continuing need for credit, cause these debts to spiral out of control. The average outstanding debt to non-institutional lenders for small holdings is NR 5,000 (US$90.00). This is a huge amount when one considers that the average annual income for such farms is just NR 6,300 (US$113.00).[14]

In 1994, only about seven per cent of households managed to make repayments on the loans,[15] which begs the question why lenders make loans which will stand such a low chance of repayment. The answer is complex. In some situations the motive for lending may genuinely be to help a family in a crisis situation. More often there is a less altruistic motive; in the village context the creditor wields a lot of power and the debt becomes a mechanism of control.

Debt Farming

Sidhashwar is one of three small villages which make up Ward no.2 of Maharudra (VDC). It is situated in a remote part of Baitadi District, over a day's walk from the nearest road head in the Far-Western region. The village is made up exclusively of low-caste families of which there are about 140 living in the VDC as a whole. The credit requirements of the community are serviced by five money-lenders who are high-caste and live elsewhere in the valley. Interest rates are much higher on loans made to low-caste groups, commonly set at five per cent as opposed to two or three per cent for high-castes.

Bagi Ram Dyal is 30 years old. He lives together with his four brothers, their wives and children in an extended family household. Between them they have a total debt of NR 90,000 (US$1,620). He has about six *nali*[16] (0.08 hectare) of land.

> *Some years ago my father became ill with tuberculosis and we had to send him to India for treatment. It cost nearly NR 30,000 (US$540.00). We raised the money from local landlords and money lenders. You can't go to the bank for medicine. They charged me five per cent interest a month but I had no other option or my father would have died. At the moment we owe*

NR 90,000 (US$1,620) *and NR 40,000* (US$720.00) *of that is just from the interest. I am lucky, I am working now as a driver in India and get NR 4,800 (US$86.00) a month and so I am able to send money home every three or four months to pay the interest. Two of my brothers are also working in India but even together we can hardly cover the interest payments.*

Another agricultural labourer, Gora Dayal Lohar, says he has a debt of NR 150,000 (US$2,700) but is interrupted by other village members who remind him that the debt has risen to NR 250,000 (US$4,500) in total. For Gora Dayal Lohar the amount is academic. He has been married three times, his two previous wives having died. He took on the debt to pay for their medical expenses.

*About NR 90,000 (US$1,620) is debt – the rest is interest. We were not told that the interest would be added on to the debt each year (*compound interest*) and there is no way to pay back the money. Once I pay the interest I have to get another loan to get food. We used to have very good land but now the money-lender uses it instead of the interest on part of the debt. At first the debt wasn't written down but when it became a lot the money-lender made a document.*

One family lost its land completely to the money-lender and was forced to move away. No one has heard of them since. But as Poran Ram Dyal, a local primary school teacher living in the same village, explains:

*It is the same with most people in the [*low-caste*] village. Every household has a debt of a minimum of NR 10,000 (US$180.00). Usually, the money-lender doesn't take the land completely, but he lets us keep just enough to survive and takes whatever else he likes. Sometimes they will come and take the best cattle, or if they need fodder they will come and cut down our crops early to feed to their cows. Now we only keep bad quality animals because otherwise the money-lender will take them.*

If they hear that someone has returned from working in India they will come and poke around to see what they have brought back and they take the best things. In social places or meetings in public we cannot say anything against the money-lender or he will humiliate us and put pressure on us again.

The worst thing is that no matter how hard we work we never see the profit. All the income from our labour goes straight to the money-lender.[17]

Discrimination and Division

It is in this situation of division and exploitation that debt bondage and slavery occur. The prevailing social structure allows a minority to exploit the poverty of the rural majority and to keep people in a dependent position. As we shall now see, this exploitation is also reinforced by Nepal's caste system.

NOTES

1 Ram Bahadur, K.C, 1986. *Land Reform in Nepal*, p2, His Majesty's Government of Nepal, USAID, GTZ & Winrock.

2 The remaining six per cent of land in 1952 was held either in religious trusts and known as *Guthi* land (two per cent), or was given over to communal ownership in the form of *Kipat* land, traditional lands assigned to some of the indigenous groups.

3 Ram Bahadur, K.C, 1986. p 3, *op. cit.*

4 *ibid.*, p12.

5 Oxfam, 1990. *op.cit.*

6 Extrapolated from Central Bureau of Statistics, 1994. *National Sample Census of Agriculture, Nepal 1991/92*, His Majesty's Government National Planning Commission Secretariat, Kathmandu.

7 Oxfam, 1990. *op.cit.*

8 Extrapolated from Central Bureau of Statistics, 1994, *op.cit.*

9 Most families with less than one hectare of land are found to be living below the poverty line. The poverty line as defined by the National Planning Commission of Nepal is the amount of income required to produce adequate calories (2256) for an average Nepali to remain active.

10 Central Bureau of Statistics, 1994, *National Sample Census of Agriculture Nepal 1991/92: Analysis of Results*, His Majesty's Government National Planning Commission Secretariat, Kathmandu.

11 Interview with Mr Kesab Badal, MP, Chair of Land Reform Commission and Dr Shiva Sharma of APROSC. ASI/INSEC September 1995.

12 Nepal Rastra Bank, 1994. pp 55-89, *op.cit.*

13 From research carried out by ASI/INSEC October 1995 – February 1996.

14 Nepal Rastra Bank, 1994, p 48, *op.cit.*

15 *ibid.*, p 71.

16 See Appendix 1 for conversions.

17 Group interview conducted in Maharudra VDC, Baitadi District, by ASI/INSEC, October 1995.

CHAPTER 7

Caste, Ethnic Group and Untouchability

We both have the same blood and same God, why do we have to face this discrimination and humiliation just because we were born into a particular caste?
Ram Bahadur Damai, Hemanta VDC, Bajhang District,
Far-Western Nepal[1]

The caste system in Nepal is bound up with the country's political structures to give religious and (until recently) legal sanction to social division and oppression. The high-caste *Brahmin* and *Chhetri* are generally dominant in most areas of society. In the rural areas they hold the best land, the senior positions on the Village Development Council, and local government posts.

The lower castes are marginalised from this society. Often this is reflected in the physical location of their settlements on the outskirts of the village or in small hamlets, in separate wards, totally removed from the main village. 'Untouchability' is widely practised in the rural areas and at best the low-caste groups have only token representation in the local decision-making process.

Caste is an extremely important factor behind the exploitation of agricultural labour. It is crucial to an understanding of the conditions which allow this exploitation to continue and at the same time defines the various groups involved. It is nonetheless an extremely complex system. Although caste is often described as rigid and unchanging, in fact it is evolving all the time. In order to see how it affects present society it is necessary to understand something about its development in Nepal.

The Evolution of the Caste System and Untouchability

The Hindu caste system is thought to have developed in South Asia towards the end of the Vedic Period (about 3,000 years ago). This was essentially a division of labour into four broad groups known as *Varna*. It was hierarchical in nature with *Brahmins* at the top, engaged in philosophical and religious pursuits, *Kshatriya* (known in Nepal as *Chhetri*), involved in the military and political field, *Vaiysha*, the businessmen and farmers, and *Shudra*, the lowest *Varna*, engaged in manual labour and traditional crafts. Under each of the *Varna* there was a complex sub-set of castes. At first there was a degree of mobility between the major caste divisions, but later it became impossible, in principle, to change your caste status from birth.

As the process of 'Hinduisation' continued within the Indian sub-Continent, the ranks of the *Shudra* castes were swelled by the vanquished and prisoners of war. This group was to be despised and many of its members had a status approximating to that of slaves. In this way, a further division gradually evolved within the *Shudra* of those whose touch was defiling and those whose touch was not. The 'untouchable' group was effectively expelled from society, was forbidden to enter the temples, and prohibited from any contact with high-caste groups by religious texts. These texts also prescribed the punishments and purification procedures to be followed should the caste rules be violated. According to these writings some 'untouchable' members are more polluting than others. The very act of looking at people belonging to certain 'untouchable' groups is considered polluting for a high-caste person who must be purified by looking directly into the sun.

The classical caste system and practice of untouchability is subject to many regional variations within South Asia; this is particularly true of Nepal.

Official figures state that 87 per cent of the population of Nepal is Hindu.[2] This statistic, however, reflects an interpretation of Hinduism which is much broader than the classical Brahminism of the scriptures. Hinduism in Nepal evolved in a piecemeal fashion and was never imposed as a coherent whole by a foreign conquering power. It has an indigenous strain built around the worship of Pashupatinath[3] in the Kathmandu Valley, and this has developed in parallel with more orthodox forms brought to the Himalayas by migrant *Brahmins* from the south. As migrants, these groups were not always in a position to impose an orthodox caste system on an already stratified society. In the early period at least, this gave rise to a hybrid form in the host society which reflected the relative power of the various ethnic groups at that time. Over the years, many of the indigenous peoples of Nepal have introduced Hindu gods into their own belief system, and by the same token, the *Brahmin* groups have allowed some religious practices derived from the Shamanistic and Animistic traditions of the pre-Hindu inhabitants to be incorporated into Nepalese Hinduism. It has also been greatly influenced by Buddhism, Nepal's second major religion.

Like the development of Hinduism, the caste system in Nepal is similarly a complex affair. Being at once political, as well as religious in nature, it is as much a product of the country's various historical power struggles, as it is of this composite form of Hinduism. Furthermore it has had to incorporate in its design the region's many different ethnic groups.

At times in the country's history, the caste system has been used by a particular ruler or regime to cement its own power base, while at other periods it has been less significant. Where it did exist it generally did not extend much further than the immediate seat of power. This gave rise to the development of three separate regional caste-based communities in Nepal: one around the Kingdom of the *Khas* (Nepali-speaking) in the western hills, another around the *Newar* people (Newari-speaking) of the Kathmandu Valley, and another around the Maithili-speaking people in the *terai*. It was in fact only with the *Muluki Ain* (Civil Code) of 1854 that a comprehensive caste system was codified and systematically applied across the country by the first Rana Prime Minister, Jung Bahadur Rana.

The *Muluki Ain* (Civil Code) 1854[4]

The *Muluki Ain* was an attempt to synthesise the three separate caste systems which had evolved in partial isolation in the hills, the Kathmandu Valley, and the *terai*. It was at the same time a legal code which prescribed punishment for transgressions of caste laws. These laws were not only designed to keep the different castes pure, but also affirmed and maintained the different caste roles. The code also provided for the differential legal treatment and sentencing of individuals on the basis of their caste. Although the legal powers of the *Muluki Ain* have since been repealed, it still provides the blueprint for the present arrangement of Nepal's major castes and ethnic groups into a hierarchical structure.

The *Muluki Ain* separated the population into two major groups, those from whom *Brahmins* could accept water, and those from whom they could not. These two categories were then sub-divided into five further groups: *Tagadhari* (Wearers of the Holy Cord), Non-enslavable *Matwali* (Matwali means Alcohol-Drinkers and is used as a term to categorise ethnic groups in Nepal), Enslavable *Matwali*, Impure but touchable castes, and Untouchable castes (see table).

Subsequent amendments to the *Muluki Ain* of 1854 did not significantly alter these caste categories until the new constitution of 1950 which declared everyone to be equal under the law. Even after this period, the authorities have struggled to come out clearly against the caste system. The legal status of the caste system was not specifically repealed until the revised *Muluki Ain* of 1963. Even then a contradictory Palace statement was subsequently issued declaring that the caste system itself had not been abolished.[5]

The same ambiguous approach has continued even since the establishment of democracy in Nepal in 1990. The new constitution seems

Hierarchy of Major Caste Ethnic Groups [6]

	Caste Divisions	Castes and Ethnic Groups
Water Acceptable	Tagadhari (Wearers of the Holy Cord)	Brahmins, Thakuri Chhetri, Chhetri, Newar Brahmins and other high Newar castes
	Non-enslavable Matwali	Magar, Gurung, Limbu and some other Newar castes
	Enslavable Matwali	Sherpa, Tamang, Chepang, Kumal, Hayu, Tharu, Gharti (descendants from freed slaves), Satar, Dhimal, and Dhangar
Water Not Acceptable	Impure but touchable	Mainly consisting of Newar occupational castes, Muslims and Europeans
	Untouchable	Hills: Kami and Lohar (blacksmiths), Sarki (leather workers and cobblers), Damai (tailors and musicians), Badi (dancers), Gaine (travelling minstrels) *Terai*: Dhobi (washermen), Halkhor and Chamar (latrine cleaners), Dushad and Dom (undertakers and sweepers), Tatmas and Musahar* (labourers and ditch diggers) Kathmandu Valley: (Newar)Pore (fishermen) Cyme (scavengers) Pode (sweepers) Kusle (musicians)

* Musahar are considered an ethnic tribal group but are also treated as one of the lowest forms of 'untouchable' in the *terai*.

explicit, stating:

> *No person shall be discriminated as untouchable or debarred from attending public places or using things of public utility on the basis of caste. Such an act shall be punishable by law.*

However a further provision in the *Muluki Ain* of 1991 completely undermined this strong statement by allowing discrimination which had existed 'traditionally'.[7]

The effects of the 1854 *Muluki Ain* on Nepali society as a whole have lasted long after the law itself was repealed. Firstly, it has reinforced an hereditary elite among the *Brahmin*, *Chhetri* and *Newari* high-castes, which continues to dominate most aspects of economic and social life. This status quo is maintained through a process of extended nepotism known as *Afno Manche* or "one's own people" where jobs or political favours will be granted first to the extended family and friends, regardless of their qualifications and ability. Secondly it has had a profound effect upon how the ethnic groups are organised.

Ethnic Groups

The enforcement of the caste system under Jang Bahadur Rana's Civil Code represented the final ascendency of the Nepali-speaking hill people, the descendants of the *Khas* and Hindu migrants, over the other ethnic groups of the area. This group, known collectively as the *Pahariya* (meaning hill people), has subsequently become the most numerous group in Nepal and as such is the majority society. However, the division of Nepal's many ethnic groups into two categories under the Civil Code reflected the relatively high status of the ethnic groups from the middle hills, which participated in the unification of Nepal under King Prithvinarayan Shah. A lower status of 'enslavable *Matwali*' was given to those groups who occupied the lowland areas or the high eastern hills. Evidently this status indicates that members of such groups could be made into slaves as a legal punishment. These many different ethnic groups, some with populations of just a few thousand, came into contact with the majority society, and Hindu religion, at a later date and were categorised as primitive by Pahariya society. It was from these groups that many slaves were recruited for work on public building projects.

The present hierarchy continues to reflect this division and the 'enslavable *Matwali*', such as *Tharu*, *Chepang*, *Satar*, *Dhimal*, and *Dhangar* in the *terai*, are much more exposed to debt bondage than the higher status *Matwali*. The latter, on the other hand, are relatively mobile socially and better represented in both the political and business spheres.

The lines between occupational caste and ethnic group are very often blurred and this is most apparent in the case of some of the *terai* tribal groups. For instance, some anthropologists define the *Musahar* of the eastern *terai* as an 'untouchable' occupational caste charged with ditch-digging and agricultural work.[8] Others maintain that they are a tribal people who have migrated into eastern Nepal from Bihar and West Bengal.[9] Exactly the same applies to the *Satar*, and *Dhangar*.

Perhaps the most significant aspect of the caste system relating to bonded labour in the rural areas, is the propagation of the system of 'untouchability'.

'Untouchable' Groups

Quite apart from the different ethnic groups in Nepal which are vulnerable to debt bondage, approximately 15 per cent of the population (almost two million people) belong to 'untouchable' castes.[10] The different 'untouchable' groups stem from the three different caste systems which evolved in the western hills, the Kathmandu Valley and the *terai*.

The *Kami* (known as Lohar in the Far-West), *Damai* and *Sarki*, although originally from the hills, are now found all over the country. Together these groups account for 8.5 per cent of the total population of Nepal.

In the *terai*, there are occupational castes such as *Halkhor* and *Chamar*, who clean the latrines, *Dushad* and *Dom* who deal with funeral preparations and dispose of carcasses, and *Tatmas* who are involved in agricultural work.

The *Newar* occupational castes have a further set of divisions but are found mainly in urban areas (concentrated in Kathmandu), so are rarely involved in agricultural work outside the Kathmandu Valley.

A dwindling number of people from 'untouchable' occupational castes continue to practise as blacksmiths, tailors, and leather workers in the rural areas. With the increasing availability of imports and mass produced items, they can no longer rely on their traditional skills for their income. Most have only a small piece of land for a kitchen garden and so are forced to work as *haliya* on other peoples' land. Their role as *haliya* is defined by their caste. The caste system discriminates against them and keeps them poor, dependent and disempowered. As such it becomes the backbone which supports bonded labour in the rural areas.

'Untouchability' and Caste Discrimination

In the villages low-caste men and women are not allowed inside high-caste houses and in some areas offices and tea shops are also out of bounds to them. Likewise low-caste members are barred from entering most of the temples freely[11], and there is a continuing system of separate drinking taps in many villages.

During 1993, INSEC conducted a survey and found incidences of caste discrimination occurring in 43 of Nepal's 75 districts. Sometimes extreme violence was reported.

In Sindupalchowk District, three year-old Laxmi Bishnukorma was beaten by P Bhattarai when she unwittingly touched the local well while playing. The next day her body was found floating in the same well. The death was reported as an accident even though a few days later Nav Bahadur Bishnu Karma (another untouchable man) was beaten up for drinking from the same well and told "One person has already lost their life, do others intend to die?"[12]

There is a greater tendency towards caste-based discrimination in the west of Nepal than in the east. In a survey conducted for the Save the Children Fund US (SCF US),[13] 90 per cent and 94 per cent of respondents from 'untouchable' castes in two western districts reported such discrimination. By comparison the figure for Jhapa District in the Eastern region was much lower at 69 per cent. This supports the observations made in the course of this research that bonded labour was more pronounced in the *Haliya* system in the western hills among caste-based communities.

Dependency

The caste system keeps lower castes poor and therefore dependent. The same SCF US survey concludes that caste discrimination is one major reason why 'untouchable' groups remain poor. This ensures that there is always a group of destitute agricultural labourers who can be easily exploited.

'Untouchable' communities are forced to live in separate settlements completely removed from people of high-castes. In spite of their poverty, these settlements are a low priority for village development projects, as it is the high-castes which hold power locally and control where and how development funds are spent. Even in relatively rich areas these communities can remain untouched and in abject poverty. For instance, in Ward no. 2 of Arjundhara Village Development Committee in Jhapa District there is a settlement of 'untouchable' families called Ekant Tole. While the main village has been supplied with electricity for the last 20 years, it has not yet reached the short distance to Ekant Tole. No roads have been constructed and the settlement has no proper supply of drinking water.[14]

The problem of access to drinking water for 'untouchable' members is one which is repeated all over the country. If an 'untouchable' drinks from the village tap, that source is considered to be polluted for the high-castes. This leads to a system of separate taps in public places and very often 'untouchable' members are completely banned from using the best water sources. They must therefore rely on less clean water supplies with all the attendant health risks. One study of a village north of Pokhara concluded that there was a much higher rate of gastro-intestinal disease among 'untouchable' groups as a result.[15]

'Untouchable' castes are prevented from doing other forms of work. They are expected to work in their traditional menial roles and any attempt to work outside them is resisted. Ninety-seven per cent of respondents to SCF's survey of 'untouchable' members reported that they were working in their traditional caste occupation or in menial labouring roles. The very fact of being an 'untouchable' in itself limits the possibilities of generating other income. The level of informal debt among 'untouchable' groups is often very high. One reason is that interest rates from money-lenders are often much higher for 'untouchable' groups than they are for high-caste. In three VDCs in Baitadi District, 'untouchable' groups were charged five to six per cent interest per month on loans from money-lenders while high-caste people were commonly charged two to three per cent.[16]

Institutionalised Discrimination and Disempowerment

'Untouchable' groups are effectively denied the means of improving their situation or asserting their rights. This starts with education. A total of 88 per cent of 'untouchable' group members identified in the 1991 Census had never been to school. This compares with a national average of 68 per cent.[17]

Only 0.2 per cent of people with 'untouchable' status are employed in a professional capacity; this compares with a figure of 0.7 per cent for the population as a whole.[18] The police force and armed services employ very few members of 'untouchable' castes and when they do they are given very low positions. In the Royal Nepal Army 'untouchables' are not allowed to enter the main barracks compound and have to live outside. Dahrendra Bahadur Rasaila, from Bajhang District, was declared to be ineligible for a police job on the grounds of his 'untouchable' caste in 1993.[19] Discrimination is widely practised in the civil administration. During the land survey carried out in Bajhang District in 1992-93, government surveyors did not register land in the possession of the Sarki community which had been using it for generations, but instead registered it illegally in the names of local headmen.[20]

Based on government statistics, which do not give a comprehensive breakdown by caste, an estimated 10 to 15 per cent of the population is considered 'untouchable'. In 1975 not one 'untouchable' was to be found among the district *Panchayat* chairmen and vice-chairmen.[21]

Jeri Ram Lohar, 33, works as a blacksmith in Ward no. 9 of Melauli VDC, Baitadi District in Far-Western Nepal. Three of his six children go to the local Durga Primary School.

We are looked down on very badly. There are very few opportunities for us. I want to educate my children so that they can stand on their own two feet and manage their lives properly.

I hope that one day they will go to the high school, but who can tell what will happen?

Jeri Ram has a debt of NR 7,000 (US$126.00) on which he only just makes the interest payments of five per cent each month. Any unforseen circumstances could instantly jeopardise his children's chance of education. Even at school his children face caste discrimination.

There are separate taps at the school for the 'untouchable' children and they treat them differently. During* Soshati Puja (a religious festival) *last year, the staff were giving* tika (coloured spot placed in the middle of the forehead*) to the children. They separated the 'untouchable' children and gave* tika *to the high-castes first. Then the teachers gave the materials to the low-caste children and told them to do it themselves! [22]

If Jeri Ram succeeds in his wish to get his children through high school, they will be part of only 0.4 per cent of 'untouchable' members who achieve a School Leaving Certificate (SLC). This compares with a national average of 2.8 per cent (i.e. seven times higher).[23] For those who do pass the SLC it is only the first in a series of hurdles.

Their representation in the present political structure remains very low with only 1.1 per cent of parliamentary seats being occupied by members of 'untouchable' 'groups in 1991 and no cabinet ministers.[24]

This discrimination is not found only in political parties which support the status quo, but is also a feature of the communist CPN UML.

'Untouchables' are unrepresented and are unable to assert their rights. This leaves them highly vulnerable to bonded labour. The majority of 'untouchable' people are forced to remain silent in the face of discrimination even if they perceive it as unjust.

A significant proportion say that they consider it to be their fate in this life. Most 'untouchable' members believe that there is little point in reporting caste discrimination or seeking legal redress as the odds seem stacked against them. The vast majority of Nepali lawyers and the judiciary come from high-castes. The SCF US survey identified 31 legal actions involving caste-based discrimination but noted that only seven of these had found in favour of the 'untouchable'.[25]

Caste and Bonded Labour

The previous chapter explained how the system of land holding in Nepal's agrarian society leads to bonded labour. It is the caste system which locks the different groups into this structure. It creates a cycle of dependency and disempowerment for lower caste members which keeps them from asserting their basic rights and freedoms. In this respect, it has proved stronger than the law.

The consequence is that hundreds of thousands of people are forced into a servile status.

NOTES

1 Mishra, S, 1995, unpublished research, INSEC, Kathmandu.

2 Central Bureau of Statistics, 1995, *Population Monograph of Nepal,* His Majesty's Government National Planning Commission Secretariat, Kathmandu.

3 The earliest form of Hinduism in Nepal is thought to have grown up around the worship of the god Pashupatinath (another name for Shiva) in the Kathmandu Valley.

4 A comprehensive discussion of the *Muluki Ain* is provided in Hofer, A, 1979. *The Caste Hierarchy and State in Nepal: A Study of the* Muluki Ain *of 1854,* Universitätsverlag Wagner, Transbrück.

5 Hofer, A, 1979. p 204, *op.cit.*

6 This table is included as a guide only; it is not intended to be comprehensive.

7 INSEC, 1994, *Human Rights Year Book 1993,* pp 349-379, INSEC, Kathmandu.

8 Bista, D.B, 1967. p 113, *op. cit.*

9 Gautam, R. and Thapa-Magar, A.K, 1994. *Tribal Ethnography of Nepal Volume II*, p 83, Book Faith India, Delhi. *op.cit.*

10 Hofer, A, 1979. p 205, *op.cit.*

11 Cox, T, 1994. *The Current Socio-economic Status of Untouchables in Nepal*, p 95, Occasional Papers in Sociology and Anthropology, Vol. 4, Tribhuvan University, Kathmandu.

12 INSEC, 1994. p 370, *op.cit*.

13 Cox, T, 1994. pp 90-109, *op.cit*.

14 INSEC, 1994. p 372, *op.cit*.

15 Macfarlane, A, 1981. "Death, Disease and Curing in a Himalayan Village" in *Asian Highland Societies in Anthropological Perspective*, C. von Furer Haimendorf,(Ed.), New

Delhi. Cited in Cox. T, 1994, p 95, *op. cit.*,

16 From research carried out by ASI/INSEC, October 1995.

17 Extrapolated from Central Bureau of Statistics, 1993. *National Sample Census of Agriculture Nepal 1991/92*, His Majesty's Government National Planning Commission Secretariat, Kathmandu.

18 As above these figures are intended for comparison only. The total number of individuals from the untouchable groups employed in professional or technical work is 4,956.

19 INSEC, 1994. p 361, *op.cit.*

20 INSEC, 1994. p 363, *op.cit.*

21 Hofer, A, 1979. p 205, *op.cit* .

22 From research carried out by ASI/INSEC, October 1995

23 The 1991 Census does not identify all untouchable groups. The figures quoted here are based on those untouchable groups which the Census does clearly identify. The number who have passed SLC and above (8,023) is expressed as a percentage of the total population of untouchable groups (2,180,547). The total figure obviously includes children who could not yet be expected to pass. The percentage is therefore only meant to serve as a comparison with the national average which was calculated in the same way.

24 Gurung, G.M, 1994. *Indigenous Peoples: Mobilisation and Change,* p 145, HISI Press, Kathmandu.

25 Cox, T, 1994. pp 97-98, op.cit .

Part III
Prospects for Change

CHAPTER 8

Debt Bondage in the South Asian Context

Although the practice of debt bondage has not been widely investigated in Nepal, in India and Pakistan it has received considerable attention from academics and campaigners alike. This represents a huge bank of knowledge which can be applied to Nepal. There are striking similarities between the systems of bonded labour which have been uncovered in Nepal and those which have been known about for many years in India and Pakistan. There is even a similar nomenclature. This is hardly surprising given the shared linguistic and cultural traditions and the historic movement of different peoples across the region.

The roots of bonded labour go back much further than the present geo-political boundaries in South Asia. They pre-date the advent of Buddhism and the introduction of Islam and can be traced back to the early Aryans[1].The origin of bonded labour is linked with the hierarchical caste system and the creation of servile castes by the ancient *Brahminic* law-givers. The practice continues to affect almost exclusively low-caste or 'tribal' groups. This is true even in Pakistan where Hindu 'scheduled' castes are badly affected.

India

There are many different terms used in India to describe what are essentially subtle variations of the same system of bonded labour. These vary from state to state and according to the different tribes or castes which are involved. Bonded labour has been practised for centuries and the words used in 1858 to describe the system of *kamiauti* in Bihar, still hold true today:

> *You will find it nearly all over the country, I believe, in one form or another... A man borrowed money and gave a bond to work it out binding himself to work for the lender, whenever he was required, without pay. The lender of course, required his services at harvest and the other busy seasons of the year, when the debtor could have got work and pay elsewhere; and when work was slack, the lender, of course, did not require his slave's services. He*

could make nothing elsewhere; all he got when working was food, and sometimes a bit of cloth once a year... The debtor could never work out his debt... The son, daughter or other nearest relation of the debtor used, in case of his death, to be considered liable...[2]

The same system operates throughout Bihar today and is known as *kamiauti* or *kamiah, harwahi* or *kandh.* In other states it is known by these or other names including *halia* in Orissa, *harwashee* or *kamia* in Madhya Pradesh, *hali* in Gujarat, and *haris* in Uttar Pradesh and Maharashtra. In many cases it is obvious that these variations simply reflect different dialects or different transliterations of the same word.

Because the system is so old and the status of bondage handed down from generation to generation, it is often difficult to establish how the bonded labourers originally get into debt. Social pressures demand large expenditure, even in impoverished families, at times of marriage, birth and death. Often these expenses can only be met by arranging a loan at high interest from the local landowner. Debt is a daily fact of life for the large majority of agricultural labourers in India and is used to pay for medical expenses or just to get enough food into the house. A survey in Bihar published in 1976 indicated that 72 per cent of agricultural labour households were in debt.[3]

Although bonded labour is most common in areas of India which are particularly impoverished, it is simplistic to suggest that poverty is the only or most important cause. Bonded labour depends on a division of the rural population into a class of landless and oppressed agricultural workers on the one hand, and an entrenched, landed class on the other. In India this division is neatly kept in place by the constraints of the caste system. It is noted that the highest incidence of bonded labour is found in areas which are dominated by powerful upper-caste Hindus, and where a rigid caste system is enforced.[4]

While 'untouchable' castes are severely affected by bonded labour, the other major group affected are *adivasi* or 'tribals', the word given to India's indigenous peoples. In the same way that bonded labour is associated with a rigid caste system, it is also reported to be high where there are large numbers of people from *adivasi* groups.[5]

In the case of *adivasis*, debt bondage is the end result of a process of land invasion and resource alienation which has been going on for centuries. This is very similar to the situation of the Tharu in Nepal who fell victim to the *Kamaiya* system. In the Indian state of Maharashtra, for example, the Gujar settlers were moved into indigenous lands in Dhule District, originally at the instigation of the British, to work on cotton plantations. The Bhil people, indigenous to the area, gradually saw their land deforested and turned over to sugar cane and cotton production.

Soon they were left with little option but to work for the settlers as bonded labourers and ended up living in huts on the landlords' lands and carrying out all aspects of the agricultural work. They were treated abominably; there were reports of flogging, rape and the burning of the Bhils' homes in the 1970s.[6]

There have been attempts at various stages in India's history to tackle the problem. These include the Debt Bondage Abolition Regulation Act of Bihar and Orissa of 1920, and similar legislation passed in Madras in 1940. By far the most comprehensive attempt to address bonded labour, however, was the Bonded Labour System (Abolition) Act, introduced by Indira Gandhi's government in 1976.

The Act proclaimed all bonded labourers free from their debts, provided for their social and economic rehabilitation, and specified punishment for those who continued to make use of bonded labour. It also set up a system to implement the Act which put considerable powers and responsibilities in the hands of the District Magistrate. The District Magistrate, the senior administrative official in each district, was to be supported in this work by local Vigilance Committees made up of a representative cross-section of the local community.

Although the Act is a worthy piece of legislation, 20 years after it was passed it is clear that it failed in its main objective of abolishing bonded labour. The extent of this failure became obvious even as early as 1981 when the Gandhi Peace Foundation conducted its major statewise survey of bonded labour.[7] This concluded that in just 10 of India's 22 states there remained a total of 2.6 million bonded labourers.

Campaigners did not have to look far to understand what was going wrong. It was clear that at local level there was little enthusiasm for implementing the Act, either on the part of the already overloaded District Magistrates or the local Vigilance Committees. A study in Santhal Parganas in Bihar published in 1984 tells a story which can be repeated for most states.

> *The role of the advisory committees at the district and sub-divisional levels can be well assessed by the fact that the District Vigilance Committee, since its inception in 1977, could meet only twice until the end of 1982. The Sub-divisional Advisory Committee of Deoghar had met only once during this period and in all the other sub-divisions of the districts the relevant committees have not been able to hold any meetings at all.*[8]

The main instances where significant numbers of bonded labourers have been successfully emancipated in India have been where local social action groups have been involved. There are many examples of where NGOs have successfully identified bonded labourers (where the government had not), and either forced the local administration to take action under the Act, or bypassed the local level altogether. The typical way in which NGOs have done this is by filing public interest litigation with the Supreme Court of India on behalf of the bonded labourers identified.

The first such ground-breaking case was brought by the Bonded Liberation Front (BLF)[9] in 1984. The BLF conducted a survey of the working conditions in Faridabad stone quarries in Haryana State and addressed a letter to the Supreme Court alleging the existence of bonded labour. After confirming the truth of the allegation the Court acted to free the workers concerned and have them returned to the states from which they had been taken. A follow-up case brought by the journalist Neeraja Choudhary obliged the State Government of Madhya Pradesh to rehabilitate them.

Similar actions have been taken by other social activists. In Madhya Pradesh, for example, activists from Raipur District worked with the Supreme Court to bring about the release of 4,000 bonded labourers in 1988. The story is repeated in Punjab, Maharashtra and Rajasthan among other states.

Through the actions of NGOs with the support of the Supreme Court, and, in certain states, the active participation of senior state officials, some 200,000 bonded labourers were 'released' by 1990.[10] However, this amounts to just 10 per cent of the bonded labourers identified in the Gandhi Peace Foundation survey of 1984. Bringing about full rehabilitation under the Act has proved even more difficult and bringing prosecutions harder still. By 1991 only 773 keepers of bonded labour had been arrested and even fewer were prosecuted.[11] In March 1995 the Supreme Court ordered a series of investigations to be carried out in 13 different states which indicated that, as suspected, a massive problem of bonded labour remained.[12]

Most recently, in 1996 a Supreme Court Commission reported on the extent of the problem of bonded labour in Tamil Nadu, estimating there were one million bonded labourers in the State. The report concluded that bonded labour existed in substantial numbers in all 23 districts of the State, and listed 20 occupations where it was rife including the industrial sector, as well as traditional occupations such as farming and stone quarries.

Pakistan

A law banning bonded labour was adopted in Pakistan in 1992, although it was not until 1995 that the Ministry of Labour issued the Bonded Labour System (Abolition) Rules which provided the necessary instructions for local government officials to implement the Act. These reforms have come after many hard years of campaigning by a handful of anti-bonded labour activists. All the indications are that it will be many years before the law passed five years ago is implemented properly.

The 1992 Act came about largely as a result of campaigns against bonded labour in the brick kiln industry, particularly in Punjab. Two NGOs, the Bonded Labour Liberation Front (BLLF) and the Human Rights Commission of Pakistan (HRCP) played leading roles in forcing the legislation through. The BLLF has campaigned against child bonded labour in the hand knotted carpet industry, while the HRCP has focused more on adult bonded labour in the agricultural economy of parts of Sindh Province. The murder in 1995 of Iqbal Masih, a child campaigner with BLLF, received unprecedented international media attention and forced the issue onto the centre stage of Pakistani politics.

The issue of direct interest for the purpose of this report, however, concerns the shocking reports of the widespread use of bonded labour in agricultural districts of Sindh.

In Sindh, bonded labour mainly affects *haris*, a term describing landless farm-workers who live and labour on their landlords' farms either on a share-cropping basis or for payment in cash or grain. In southern Sindh, where bonded labour is most notorious, *haris* are mainly drawn from formerly pastoral Hindu tribes concentrated in the district of Thar Parkar, the eastern-most district of Sindh bordering the Indian state of Rajasthan. They are from the lowest 'untouchable' castes including Bheels, Kohils, and Meghwars. In upper Sindh, *haris* are predominantly Muslim and there is less evidence of bonded labour. According to the HRCP, the majority of the *haris* in lower Sindh are not registered as citizens and are thus denied tenancy rights.[13]

There has always been gross exploitation of agricultural labour in Pakistan, very similar to the sorts of bonded labour found in other areas of South Asia. Recently, however, a new and extremely vicious form of bonded labour has been reported, with workers and their families being kept in 'private jails'.

In 1991 a detachment of the Pakistan Army raided a 'private jail' owned by Haji Chulam Kokhar in Tando Allahyar and freed 295 *haris* held there. The families had been kept under guard, there were reports of women being raped and beaten indiscriminately, and some workers were made to wear fetters and chains when they went to work in the fields so that they would not escape.

Activists from the HRCP carried out further investigations which revealed that this was just one of scores of similar private jails operating throughout the Province. These walled compounds, sometimes equipped with electrified fences, are used to confine bonded labourers for work on the landlords' estates.

The descent into debt bondage is almost identical to the experience in other parts of South Asia. A survey conducted in 1979 by the Water and Power Development Authority (WAPDA) revealed that 51 per cent of peasant farmers in Pakistan relied on loans to meet production costs and that 91 per cent of them turned to their landlord as their main source of credit. Banks will not lend to pay for marriages or other festivals and thus, for so-called 'social credit' *haris* are totally dependent on their landlords.

However, the practice of loaning money is not the problem. The problem lies in the dishonesty of the landlords and their agents, and the fact that the uneducated and marginalised *haris* are easily exploited. In the informal sector loans are not regulated and very often usurious interest rates are charged. In addition, landlords charge the share-croppers working their land for seed grain, even though this is illegal under Pakistan's tenancy laws. The cost is usually added to the share-cropper's loans. The dues are collected every harvest and the peasant farmer quickly descends into a situation of debt bondage where he is working only to pay the interest on a loan. The landlord will not release him until the loan is repaid in full. To stop the bonded labourers absconding, they are kept in the private jails. Like the *Kamaiya* system in Nepal, when the landlords need cash they may 'sell' their *haris* to neighbouring land owners. Here the buyer agrees an amount with the seller ostensibly to clear the loan taken on by the *hari* allowing him to leave his original landlord. The amount is then added to what the *hari* owes his new landlord.

The landlords in Sindh are some of the most powerful individuals in Pakistan and are often well able to resist attempts to liberate the *haris* on their land. Their control often extends over the local police and administration. The handful of district level officials who have been successful in conducting raids and freeing bonded labourers have been relocated to different areas with no explanation, leading human rights activists to believe that high level officials are trying to frustrate implementation of the Act under pressure from the landlords. One case reported in *The Frontier Post* (26 June, 1996) shows how a large landowner, Mureed Khan Maree, was able to recapture from police custody over 100 bonded labourers who had just been released in a raid on his land. No charges had been brought against the landlord for keeping bonded labourers. Pakistan's Bonded Labour System (Abolition) Act, like its Indian counterpart, relies upon Vigilance Committees to help enforce it. Only a very few Vigilance Committees have so far been set up and those which have seem to be dominated by landowners and are ineffective. These problems are compounded by the lack of awareness among district officials of their responsibilities under the Act. Shakeel Pathan, coordinator of the HRCP's Bonded Labour Task Force, related in an interview conducted for *The Herald* in June 1996 how the Task Force had written to deputy commissioners and senior police officials about the issue but found that many simply did not know about the Act:

> *We literally made hundreds of copies of the document and distributed them among government officials, journalists, human rights activists and others.*[14]

The HRCP has recommended that drastic changes need to be made to the Vigilance Committees in order to make them effective and more representative and have carried out seminars and meetings to publicise the Bonded Labour System (Abolition) Act.

Bonded Labour as a South Asian Problem

Bonded labour has been acknowledged by the authorities in both India and Pakistan to have been a problem for centuries. It has survived all manner of political and social upheavals and changing national boundaries. It is an established part of agricultural life throughout the region and, indeed, this has been used sometimes as an argument to support retaining this form of slavery, so integrated is it into social and economic life.

Until two years ago, however, the Nepali Government implied that Nepal was the exception in the region and denied that any form of bonded labour existed in the country. It was only after overwhelming evidence was presented to the Government by NGOs in the early 1990s, that the problem was acknowledged. Even now, the official Government position is that bonded labour is a very localised problem in a remote part of the country and affects just a few thousand people belonging to one particular ethnic group. In contrast, this report shows that there is evidence of a much more widespread problem affecting large numbers of people in the rural areas of Nepal and that, in most cases, it is very similar, if not identical to the patterns of debt bondage in India and Pakistan.

The fact that Nepal is having to deal with the problem of bonded labour relatively later than its neighbours presents it with particular difficulties, but also some advantages. While there is clearly a great deal for the authorities to do to tackle the problem, the experience built up in India and Pakistan provides potential models of how to proceed, and may help accelerate progress. These options are described in Chapter 9.

A further point which becomes clear from the regional perspective is of particular importance for Nepal as it strives to develop its industrial capacity. This concerns the insidious nature of bonded labour which, if left unaddressed, infects many other industrial sectors. While the roots of bonded labour are clearly embedded in the poorest agricultural areas of the country, the branches have extended into the manufacturing and service sectors, and have infiltrated the wider economy in India and Pakistan. According to an Indian government report in 1991, evidence was found to suggest the presence of bonded labour and child bonded labour in the following industries: stone quarrying, fishing, brick production, road construction, *bidi* (local cigarette) production, tea plantations and tea processing, cashew processing, cotton and textile production, hosiery industry, carpet weaving, coir industry, match and firework production, pottery production, saw mills, stone crushing, and soap production.[15]

This has emerged as an issue of international importance in both countries, particularly when the export sector is involved. The movement of the last 10 years towards economic globalisation and liberalisation of international trading practices has led inevitably to comparison of labour conditions in different countries. This has brought into sharp focus issues of bonded and forced labour in South Asia and has directly affected valuable export industries. Indeed, the same issue has already had an impact in Nepal; when bonded child labour was uncovered in the country's carpet industry, the single most important earner of foreign currency, exports were badly damaged.

NOTES

1 Tripathy, S.N, 1989. *Bonded Labour in India*, p 4, Discovery, Delhi.

2 From the account of William Le Fleming Robinson, British District Collector in Bihar cited in: Tiwary, S.P, 1984. "Bondage in Santhal Parganas", in Patnaik, U, and Dingwaney, M, (Eds.) *Chains of Servitude* , pp 187-213, Sangam Books, Madras.

3 Bose, S.R, and Ghosh, P.P, 1976. *Agro-economic Survey of Bihar: A Pilot Study* , Patna.

4 *ibid.* p 200.

5 Ministry of Labour, 1991. *Report of the National Commission on Rural Labour,* Vol 1, p 102, Government of India, Delhi.

6 Menon, G, 1997. "Bondage: Its Forms and Persistence in Tribal India", in *Enslaved Peoples in the 1990's: Indigenous Peoples, Debt Bondage and Human Rights,* Anti-Slavery International and IWGIA, Copenhagen.

7 Marla, S, 1981. *Bonded Labour in India* , Gandhi Peace Foundation, New Delhi.

8 Tiwary, S.P, 1984. p 207, *op.cit.*

9 Reddy, Y.R, 1995. *Bonded Labour System in India,* Deep and Deep Publications, New Delhi.

10 Ministry of Labour, 1991. Annexure A, *op.cit.*

11 Shastri, L.B, 1991. *Study Group on Bonded Labour, Report of National Commission on Rural Labour,* Vol 2, Government of India, New Delhi.

12 Anti-Slavery International, 1996. *India: Court Actions Reveal Scale of Bonded Labour*, Statement to the 21st Session of the Working Group on Contemporary Forms of Slavery, Geneva.

13 Human Rights Commission of Pakistan.

14 Jafri, H. I, and Hassan, A, 1996. "The Road to Freedom", in *The Herald*, June 1996, Lahore, Pakistan.

15 Ministry of Labour, 1991. *op.cit.*

CHAPTER 9

Non-Governmental Organisation Initiatives Against Bonded Labour

'Sangatan' Organisation showing teacher pointing to Nepali word on blackboard – word says 'SANGATAN' which means 'organisation'. Extract from a lesson given to an INSEC *kamaiya* night literacy class in Nambasta VDC by a young Tharu activist Bajaru Chaudhary.

> *If we work among ourselves and we work together that is an organisation. W have a form of organisation, we are here to learn together and to unite and t raise a united voice which can reach further.*[1]

In the last 10 years there has been a growing awareness of the problems of bonded labour in Nepal even if so far this has focused exclusively on the *Kamaiya* system in part of the Far-Western *terai*. It comes largely as a result of the work of a few NGOs and individual activists. The ability of NGOs to openly campaign on such issues was boosted after the establishment of a multi-party political system in 1990 and it is since that time that campaigning against the *Kamaiya* system has really got under way.

The advent of democracy has also encouraged representatives of indigenous groups and others with low status in the caste system to be more vocal in challenging caste structures. They have had some success in this area, but generally the link has not been made between the caste system on the one hand and labour exploitation in the *Haliya* and *Kamaiya* systems on the other.

This chapter looks at the role of NGOs in tackling these problems.

The *Kamaiya* System

Two NGOs in particular, Backward Society Education (BASE) and INSEC have led the way in developing comprehensive programmes to address the problem. Both identified the high level of illiteracy among *kamaiya* (96 per cent) as a major factor in their exploitation and have introduced non-formal education (NFE) classes for *kamaiya* in Kanchanpur, Kailali, Bardiya, Banke and Dang Districts. More recently they have developed separate programmes to cope with the particular problems of *kamaiya* children.

The literacy classes have always had the wider motive of empowering the *kamaiya* to help them assert their rights. BASE coined the phrase 'Education for Transformation', while from the beginning INSEC has included both human rights awareness and organising people as components of its literacy classes.

In 1987 BASE organised a strike in two villages in Dang District: Chaukura and Baibung. The strike started in *Maghi* but lasted only one month, because the *kamaiya* involved were beginning to starve. As a result BASE made the strategic decision to avoid direct confrontation in the future and instead concentrate on long-term empowerment of the Tharu community as a whole. Programmes developed subsequently have included income generation and savings schemes, and the provision of legal support. The approach has been successful, and according to Krisharu Chaudhary, a member of Chaukura VDC and himself a former *kamaiya*, only about half of the village population now work as *kamaiya* and the level of *bigha* (in-kind remuneration) they receive has increased by 40 per cent. It is, he says, "*like the difference between heaven and hell*".

INSEC's programme, which significantly got underway after the watershed of political change in 1990, has from the start taken a more confrontational approach. Initially, eight *kamaiya* villages were selected for a 12-month class in each of the five problem districts. The classes move on to other villages in the following year. In this way INSEC has been able to build up a strong network of *kamaiya* activists and teachers to support and coordinate local action to challenge the system.

RAM KRISHNA CHAUDHARY

Ram Krishna Chaudhary, 24, is a *kamaiya* leader from Janaki Nagar in Kailali District. He managed to get out of the *Kamaiya* system in 1994. He received education as a child up to fifth grade, but after that circumstances changed and he had little choice but to start work as a *kamaiya*.

There were two of us kamaiya *on the landlord's farm. We did all the field work. The landlord didn't work; generally they don't know how to. It was very hard work. We went to sleep at 9pm but we had to get up again at midnight to start work. I was totally under the landlord's control. For all this work I got 12 sacks of grain a year; it was not even enough to survive on. I worked like that for six years.*

In 1994 I decided I had to leave. Leaving was very difficult. My sauki *[debt] was NR 2,500 (*US$45.00*). I refused to pay the debt to the landlord because he hadn't been paying me the agreed amount of* bigha. *I told him 'First you pay me my* bigha, *then I'll pay you back the* sauki.' *In May the landlord brought the police. I was arrested and taken into custody but all the INSEC activists came to support me. In the end the landlord paid me my* bigha *and I paid the debt with a loan from the Village Development Bank.*

Now I am working as a carpenter and life is still not easy but it is a very big difference from before. I feel free and independent. I am able to do work when it needs to be done. I'm not controlled by anyone else.

INSEC and I are now working together with the people in the kamaiya *villages. We have formed a committee, and we are making people aware of their rights. The situation is improving for* kamaiya *here; people are becoming aware because of the literacy programmes. They get a lot of things from the classes besides. I feel there will be abolition of this system but I'm afraid it will take time.*

I was scared and afraid for such a long time. Our forefathers were exploited because they were scared and afraid too. If we are not afraid perhaps we can stop the next generation from being kamaiya.[2]

Reforming the Landlords

Predictably, both INSEC and BASE initially encountered fierce resistance from the landlords. By involving the wider community in organising the programmes however, they have been able to overcome some of these problems. BASE's strategy has always been to address the problems of the Tharu community as a whole (16 per cent of its participants are actually working as *kamaiya*). They also invite landlords to attend their *kamaiya* meetings.

INSEC has introduced *Kamaiya* Support Committees in some of the villages where they work, made up of prominent members of the community who are not *kamaiya* but sympathise with their position. These committees include landlords as well, and in some cases the initial resistance has been converted into positive help.

There are indeed some landlords who are actively working to reform the system. Mohan Sing Rathor is a landlord in Pratapur VDC. He is university educated and returned to working as a farmer after a career in the civil service in Kathmandu.

> *I didn't like or agree with the* Kamaiya *system but when I came here there was no real alternative. When INSEC came here I was happy to work with it. I have* kamaiya *and I try to gradually help them. I advise them carefully of their financial position. I paid NR 10,000* (US$180.00) *for one* kamaiya *and I worked out with him how he could pay that off by working for me for two years. Now he is free he can work on a salary basis.*

The involvement of landlords in any solution to the problem is essential. Not only do they remain a very powerful group locally and nationally but they also have important insights into the practicalities of reforming the system which should be taken into account. As Mohan Sing Rathor continues:

> *Actually the* Kamaiya *system is part of a general problem in rural areas of disguised unemployment. Basically there is not enough work throughout the year to justify paying a labourer the going rate of NR 30-40* (US$0.54–0.72) *per day. In the* Kamaiya *system you have to keep the* kamaiya *all year round. It is not economical for a landlord to use the* Kamaiya *system without exploiting the* kamaiya. *We should have a system where* kamaiya *can work as wage labour at the peak time, receive a proper wage for their work, and then be free to work in other jobs during the rest of the year. This is not possible at the moment because the other employment opportunities are not really there.*[3]

A Model Campaign

It was just 10 years ago that BASE was set up as a tiny organisation composed of a few activists. It was only five years ago that INSEC first carried out a baseline survey of the *Kamaiya* system. Both organisations have subsequently participated in intense national and international lobbying and publicity campaigns, and at the same time carried out concerted work at the grass roots. This combined approach has been very successful. The pinnacle of success to date came at the beginning of 1996 with the launching of the Kamaiya Mukti Manch (Kamaiya Liberation Forum).

The Kamaiya Mukti Manch is mainly the fruit of INSEC's participation with local *kamaiya* groups, teachers and Tharu activists. Its governing body is made up of *kamaiya* leaders and human rights workers who have emerged through INSEC's work over the last five years. It is now formally registered as an autonomous organisation under the wing of the General Federation of Nepal Trade Unions (GEFONT). Many thousands of people, a large proportion of them *kamaiya*, flocked to hear the opening address in Nepalganj, Banke District, on 25 January 1996. Speakers included local politicians, a former Prime Minister, and Swami Agnivesh, the long time campaigner against bonded labour in India and Chair of India's Badua Mukti Morcha (Bonded Liberation Front).

No one, least of all the Kamaiya Mukti Manch, is pretending that the *Kamaiya* system is defeated. Everyone recognises that the system is still firmly entrenched in western districts of Nepal and that there is a long way to go. However, the launch of the organisation marks a major step forward and is an example of what can be achieved by NGOs over a relatively short period of time.

The very real and demonstrable effect of the NGO activities in tackling the *Kamaiya* system is one of the most significant developments to have taken place in the human rights field in Nepal in recent years. A definite change of culture is taking place in the western *terai*. A process is under way which has led to the system being openly challenged by the *kamaiya* themselves in a way which would have been impossible a decade ago.

Kamaiya Concern Group

Although this report has drawn mainly on material from INSEC and BASE there are several other important national and international organisations involved in programmes against the *Kamaiya* system. Early in 1997 these organisations decided to form the Kamaiya Concern Group with INSEC taking the coordination role for the first year. The core members of the group include: Kamaiya Liberation Forum, BASE, INSEC, GRINSO, RRN, DANIDA, Lutheran World Service, Save the Children (US), UNICEF, ActionAid Nepal, SAP Nepal, UNDP, Social Welfare Council and Plan International. The aims of the group are to share information about various strategies to abolish the *Kamaiya* system, to coordinate programmes at the regional level and identify priority areas for extension of programmes, and to form a powerful lobby to implement or change legislation for the benefit of *kamaiya*. The emergence of such an influential group of organisations with a broad range of expertise and resources is an extremely encouraging development. The Group has already registered that it is concerned about forms of slavery-like exploitation similar to *kamaiya* existing in other parts of the country.

THEN AND NOW

The effect of the NGO programmes is felt most strongly by people like Joshi Ram Chaudhary from Praktapur VDC in Kailali District Mid-Western Nepal who remembers bitterly the days before such open resistance was possible.

Joshi Ram is unsure of his age but thinks he is around 60 years old. In 1973, he became aware of how the system was gradually being changed by the local landlords and tried to intervene. He had previously worked as a *kamaiya* for 10 years.

In those days, until about 1973, there was a system of giving a plot of land to* kamaiya *as* bigha. *This was quite a good system for the* kamaiya. *But the landlords got greedy and decided to change to giving 12 sacks of rice instead of land. They got together and forced us to accept the new arrangement.

When he understood what the landlords' intention was Josi Ram contacted all the *kamaiya* in the village to warn them.

In the morning none of them went to work - only one person. All the cattle were kept inside, and the landlords were very worried and angry. The next day they came for me. There were 25 landlords. I found that I was alone. They caught me and I was accused of being a* kamaiya *leader. They said that since I organised the strike I would be charged personally for the lost production. They gave me* Jutakochala *('Garland of Shoes')*[4] *and I was publicly humiliated in front of the whole village. I had no choice but to leave the village. I alone was singled out - the rest all went back to the farmers.

Having returned to Praktapur, Joshi Ram's son Firolal has now taken up the struggle as an INSEC activist. Joshi Ram believes that outside support has been a major factor in helping to challenge the system. If people had been there to help me in 1973 I think I would have won that strike. I don't think the landlords would be able to do that now.

Action Against the *Haliya* System

Although many people now recognise that bonded labour exists in Nepal, few recognise that the problem is wider than just the *Kamaiya* system. There are very few NGOs working on bonded labour in the *Haliya* system and this report represents the first real research into the various patterns of exploitation. However, there has been some work with low-caste groups which indirectly addresses the problems of *haliya*.

The Sustainable Livelihood Forum (SLF) works exclusively with 'untouchable' groups in a growing number of villages in Baglung and Parbat Districts. SLF gains access to villagers by organising adult and child literacy classes, which provide a forum where problems can be discussed and solutions identified. The approach is low key and as far as possible avoids relying on help from foreign funding organisations. Instead, efforts are concentrated on encouraging villagers to use the various government offices and local services which are in theory available. Wherever possible the villagers themselves are involved in finding solutions to their problems and where necessary supplying both finance, labour and ideas.

The village of Salyantar is in Parbat District, two or three hours' walk up the Kali Gandaki Valley from Baglung town (Western Nepal). With SLF support, a village committee was set up by and for the low-caste community:

> *Previously, the scheduled castes* (low-castes) *as a group had no real identity in the village as it was all dominated by the Brahmins and the scheduled castes always worked as an instrument of the latter. Our presence in the village has given them some sort of confidence to organise on their own.*[5]

In the village of Silmi, further down the Valley, a women's group was set up by the villagers after SLF became involved. It has been running for two years and now has 60 members. Chandra Kami, the Chair of the committee, describes what it has been able to do in relation to the *Haliya* system:

> *The committee was formed because there was so much oppression from the upper castes. We organised a demonstration to increase the wages of the* haliya. *There was some violence. For 15 days we didn't allow the men to go to plough. The landlords threw stones at us and beat us with sticks but after 15 days they held a meeting and agreed to increase the wages.*

The increase in wages, however, resulted in an increased consumption of alcohol amongst the men. The same committee addressed that issue too:

We stopped the men from drinking and gambling. As soon as they got a little money they'd spend it on drink and would beat their wives and children. Now that is stopped. We didn't allow them to make alcohol. It is the unity that has made us strong.[6]

In addition to these examples of activities at grass roots level, there have been some moves at national level to challenge the caste system which could also indirectly affect the situation of *haliya*. In 1992 Man Bahadur Bishwakarma and Manohar Lal Bamrel filed petitions at the Supreme Court to try to annul the 1991 provision in the *Muluki Ain* (Civil Code) which allowed 'traditional' caste practices to continue (see page 44). The petitions were upheld and the *Muluki Ain* amended. The law now states that no one shall discriminate against anyone else at any temple or public place on the grounds of caste.

There are currently several organisations working to redress the situation of 'untouchables'. These reflect different political affiliations. Puran Ram Dyal, who belongs to the *Lohar* (blacksmith) caste, is a primary school teacher from a remote village in Baitadi District. He is also a member of the Society for the Liberation of Oppressed Castes, or SLOC Nepal (Nepal Uptidit Jatir Mukti Shamaj). He is active within the village helping the community to assert their rights.

We are not ashamed of our caste. We are proud of our caste. I don't want to change my caste, I want to change the behaviour of the super-castes. It is they who are ignorant, they who must be educated.

There is some change here but only a little. With democracy at least we have the right to speak out.[7]

Microfinance Income Generation, Savings and Credit Schemes

There have been some successful experiments with income generation and savings and credit schemes amongst *kamaiya* groups, most of which target women. Increasingly NGOs have turned to such schemes in order to provide alternative sources of income for bonded families, in the absence of any government-sponsored rehabilitation scheme.

BASE has experimented successfully with a savings scheme where small savings of rice are collected centrally from *kamaiya* families. The savings are topped up by matching grants from BASE, and the central fund used as a source of credit to fund income generation schemes such as goat and pig rearing. The women's group in Silmi (mentioned above), has also set up savings schemes and income generation projects, usually small-scale animal husbandry projects or developing nurseries for fruit trees.

In both examples the schemes have targeted women in the bonded labour families, and while this has meant extra work for over-burdened women, the schemes have been successful in building the confidence of the women, allowing them some independence, and a greater say in family finances and decision making. In the case of the Silmi women's committee mentioned earlier, the women have evidently been given the confidence to make impressive strides toward emancipating their families from debt bondage.

However, income and credit schemes should not be seen as an easy solution to the problem of bonded labour. Recent research from other parts of the world indicates that in most cases the poorest women are bypassed in both individual and group lending programmes or are least able to benefit because of their initial low resource base, lack of skills and lack of market contacts.[8] These findings would seem, on the surface at least, to be supported in Nepal where the main beneficiaries are families which have at least some land of their own for a kitchen garden and have relatively small debts. For the most exploited groups, living on the landlords' land and in debt, the options for income generation are extremely limited. In these circumstances the proceeds of any efforts that *haliya* make to increase their income are sabotaged because the local money-lenders or landlords immediately confiscate any extra income as interest on their debt. Another problem that makes income generation difficult for *haliya* groups is the fact that they are considered 'untouchable'. Bir Karki from Bajhang District explains:

> *We are very much discriminated against because of our caste. They use very humiliating words to us. We can't even sell our milk or ghee (*clarified butter*) and so we have to stay poor.*[9]

Income and credit schemes are usually small-scale and are designed to supplement the income of those concerned, rather than to replace their current income. Rarely do they provide a real alternative employment for agricultural labourers. This is especially true in the more remote areas, far from the development taking place around the towns and roads where there is growth in markets for fruit or livestock.

Income generation and credit and savings schemes certainly have a role to play. Perhaps their greatest potential application is in the hill areas, close to local markets, where most *haliya* families have at least some land for a garden where they can grow cash crops or keep goats or chickens. It is also the case that among *haliya* in the hills the debt bondage relationship is more clearly restricted to the male family members, and as such the female members are more likely to be allowed to keep what they produce.

Limitations of NGOs' Potential

There is little doubt that, where they operate, NGO programmes are empowering the oppressed groups concerned. In the case of the *Kamaiya* system the results have often been dramatic and give much hope for the future. Nevertheless even these programmes are limited in three main respects:

1) Limitations of Scale

NGOs are generally small organisations, indeed this is one reason for their effectiveness. Although there are a growing number in Nepal, they are nonetheless limited in the number of people they can reach, and the geographical areas they can cover. (They also vary considerably in the quality of their work.) In the five districts of Kanchanpur, Kailali, Bardiya, Banke and Dang (where the programmes to support *kamaiya* are concentrated) research has indicated that there are about 200 villages where *kamaiya* are found in large numbers. At present the combined efforts of the NGOs reach about 30 to 40 villages. BASE, currently the largest NGO in Nepal, serves approximately 18,000 Tharu with its literacy classes, of whom about 3,000 are *kamaiya.*[10] INSEC reaches roughly 1,200 *kamaiya* in its non-formal education classes each year and a further 800 *kamaiya* children. However, in the five districts (mentioned above) there are an estimated 70,000 - 110,000 people affected by the *Kamaiya* system.[11]

Outside these five districts, there are very few NGOs directly addressing slavery-related problems. Indeed the *Kamaiya* system itself is reported to exist in four other districts but is not being addressed by NGOs.

As yet the *Haliya* system is not widely recognised as a problem by most NGOs and there are, consequently, very few programmes which address this area of concern.

2) Problems of Access

Paradoxically it is those groups of bonded labourers who are most exploited who are often the most difficult groups for NGOs to reach and support. For example, while most NGO initiatives set out to involve the active participation of women, participation rates are extremely hard to maintain. One INSEC teacher from Praktapur VDC in Kailali gave an explanation as to why it was difficult to get women to come to his evening literacy classes saying: "*If both* (husband and wife) *go to the classes, who will cook the evening meal?*"

Indeed, from the woman's point of view, escaping from a situation of debt bondage to the landlord may not ultimately change her servile status. At 60 years of age Moti Lal Chaudhary from Praktapur was finally able to pay off his debt and get out of the *Kamaiya* system. Although

Moti Lal recognises his situation has improved since then, his wife Fulmati still has her day full with domestic chores and farming the kitchen garden in her own home:

> *For me life is just the same. What's the difference? I have to work here just as I had to work there. My hours and duties of work are more or less the same.*[12]

Those *kamaiya* who are able to participate in literacy classes are, by definition, freer than others. It is extremely difficult for NGOs to design programmes which will secure the meaningful involvement of *kamaiya* who have absolutely no free time; yet it is precisely this group which is most in need of help.

3) Problems of Relevance

In the most acute cases, victims of debt bondage and serfdom are so far under the control of their landlords that, even if they participate in NGO programmes, they cannot put their additional knowledge and abilities to use. Ram Kumari, a 19 year-old woman from Nambasta VDC in Kailali District, comes from a *kamaiya* family. She is in debt and lives on her landlord's land. She has been attending the INSEC night classes in the village for the last two months and she says it is interesting. She can write her name and is learning to read. Unfortunately she is probably right when she concludes:

> *I liked what the teacher said about working together and I think we should have rights but I don't think the landlord will let us.*[13]

4) Alternative Forms of Employment

Even if *kamaiya* are now more aware of their rights than they used to be, without land of their own, or alternative income generating activities which can provide them with food security throughout the year, they are very often still unable to assert their rights.

Even those *kamaiya* who as a result of NGO programmes have managed to escape the system often end up returning to it at a later date, having no other means of supporting their families.

Most NGOs are aware that there can be no permanent solution to the problem of bonded labour in Nepal without direct government intervention on a large scale to provide long-term solutions to the extreme inequality which has been fostered in rural areas. If the Government is to live up to its international obligations to guarantee the basic human rights of all its citizens, it must avoid taking the short term approach of maintaining the political and economic interests of an elite. This has profound implications for Nepal's development plans.

NOTES

1 From research carried out by ASI/INSEC, October 1995.

2 *ibid.*

3 ibid.

4 *Jutakochala* is a form of public humiliation. The closest western approximation might be the practice of tarring and feathering. Here, a garland made out of shoes is placed around the victim's neck and he is paraded around the streets to receive the insults and abuse from his neighbours, prior to being expelled from the village. It is often reserved for politicians who have promised more than they have been able to deliver.

5 Sustainable Livelihood Forum, 1993, *Annual Report I and II 1991-1992*, p 7, Kathmandu.

6 From research carrried out by ASI/INSEC, December 1995.

7 From research carried out by ASI/INSEC, October 1995.

8 Mayoux, L, 1997. *The Magic Ingredient? - Microfinance and Women's Empowerment.* Briefing paper prepared for the Micro Credit Summit, Washington, February 1997.

9 Mishra, S, 1995. *op.cit.*

10 *ibid.*

11 These estimates come from the three main surveys of *kamaiya* conducted by INSEC (1992), BASE (1994) and the Landless Settlers' Problem Resolution Commission (1995).

12 From research carried out by ASI/INSEC, October 1995.

13 *ibid.*

CHAPTER 10

Government Action

> **Mr Jaganath Acharya was Minister of Land Reform in the first Congress Government after the restoration of democracy in 1990. He was removed from office after just a few months, and newspaper reports at the time speculated that his removal was connected with the fact that he was intent on openly challenging some of the feudal patterns of landholding. He is widely respected on both sides of the political divide as a politician who carried out his ministerial duties with the utmost integrity. In so doing he came face-to-face with the powers which frustrate change in Nepal. Though his view may be bleak and heavily ironic, it must necessarily be taken seriously:**
>
> ***We politicians need cheap popularity and we need votes to form a government. Land Reform is a very popular slogan but it doesn't mean anything more than that.***
>
> ***We are still in a feudal society, there is an inferiority complex amongst the lower classes. We, the political leaders and educated people and NGO workers are from the higher echelons and will not allow the lower echelons to get free. We claim to be an independent and free country but most of the people are not. We exercise the power, and will not easily relinquish that to them.***
>
> ***'We are humanitarian', everyone gives this slogan but no one who says it really wants an egalitarian society. It takes time to change society. We need a kind of social revolution and that will take time.***[1]

In 1989, the King appointed a commission to look into the *Kamaiya* system in Far-Western Nepal. The commission concluded that there were no slaves there. When questioned by a researcher, one member of the commission commented that the Tharu *kamaiya* were:

> *ignorant, some are lazy and drunkards. If they want to better their life they should work harder.*[2]

The movement which overthrew the non-party *Panchayat* system in 1989 was motivated by the desire to see the introduction not only of a multiparty political system, but also of human rights.

Subsequent governments have consequently been at pains to demonstrate publicly that they are committed to defending human rights.

After publishing its research into the *Kamaiya* system in 1992, INSEC launched a major lobbying and awareness raising campaign. Two years later and 70 years after the declaration of the abolition of slavery in Nepal, the government of the United Marxist and Leninists (UML) officially recognised that a form of slavery still existed in the country. All main political parties are now in principle committed to the abolition of the *Kamaiya* system. The new UML and RPP coalition government is the latest to have pledged action against bonded labour.

The politicians have been painfully slow to translate these commitments into action. The main government action so far has been part of a broader initiative to address the problem of landlessness.

The Sukambasi Ayog
(Landless Settlers' Problem Resolution Commission)

The *Sukambasi Ayog* is a government-appointed commission set up to investigate problems brought about by landlessness and to recommend policies to improve matters. It has been functioning at various times under different governments including the *Panchayat* regime. In recent years it has focused on the problem of squatter settlements. In 1990, landless agricultural workers, the majority of them *haliya* or *kamaiya* from the hills and *terai*, began squatting on unused government land in an organised way. They occupied marginal land by the major rivers and the forest fringes and created their own settlements. The government, not wanting to encourage further forest encroachment and spurred on by local landowners who saw their *kamaiya* and *haliya* leaving their farms, at first acted to curtail the process and demolish the settlements. When it became clear that this did not present a long-term solution the Congress government in power at the time adopted a conciliatory policy by allocating areas of unused government land, away from the forested areas, for the squatters to build permanent settlements.

When the Communist Party of Nepal (CPN UML) came into power at the end of 1994 it saw the potential of this approach for addressing the problem of bonded labour. The *Sukambasi Ayog* was given the extended mandate of carrying out a survey of landless people in all but two of Nepal's 75 districts in order to allow them to register land on which they were squatting or to provide them with alternative land. The *Sukambasi Ayog* incorporated a separate programme to address the problem of *kamaiya*. A detailed survey was carried out which identified 5,356 *kamaiya* families who were both in debt and living on their landlords' properties.[3] This group was seen as a priority for action and a scheme costing NR 40 million (US$720,000) was designed to resettle this group of *kamaiya* in model villages in each of the five districts of Kanchanpur, Kailali, Bardiya, Banke and Dang. The *Sukambasi Ayog* ordered that in cases where no documentary evidence existed for a *kamaiya's* debts, they should not be considered binding. No compensation was to be paid to landlords.

Resham Bahadur Chaudhary was President of an organisation of landless people who squatted on the banks of the River Kanali in Bardiya District:

In our area there were about 18,000 people. The people were landless agricultural workers. Many of them were* kamaiya. *We had no cattle to plough so we had to use people.

The landlords in the district became very angry and worried because their* kamaiya *were running away to join us. They accused us of being thieves and ruffians. They said that we were manufacturing bombs, and that we came from India and were not even Nepali.

In late 1992 the government ordered the settlers to leave the area and for their houses to be demolished.

The army came from 75 districts to destroy our houses. By 6 pm they had destroyed everything with seven elephants and three bulldozers. They spent hundreds of thousands of rupees destroying our homes. That money could have been used to help solve the problem in a permanent way.

Following the evictions the leaders of the community went on hunger strike. Their plight and the clear injustice they had suffered captured the headlines. After 27 days the government changed its confrontational stance and Sailaja Acharya, Minister of Agriculture, was sent to negotiate a compromise. Through the auspices of the *Sukambasi Ayog* the squatters were granted a block of government land on which to build a village.

As Resham Bahadur proudly states:

***This new village is called Reshampur. RNN* (Rural Reconstruction and Development, an NGO working with agricultural communities) *supplied the materials and in 80 days we built 51 houses. Eighty-seven families have now received land... It was very hard but now we finally have our own piece of land and we have proved that we are not criminals.*[4]**

The *Sukambasi Ayog* was given considerable powers to allocate government land for the rehabilitation of bonded labourers and the landless. NR 10 million of the NR 40 million required for the *kamaiya* scheme was allocated. By the time the UML lost power in September 1995, less than 1,000 *kamaiya* had received land, and none of the model villages had been built. Nonetheless, almost 60,000 landless families across Nepal were given land registered in their own names.

The succeeding coalition government dissolved the *Sukambasi Ayog* and put an end to the *Kamaiya* scheme instigated by the UML. Some of the responsibilities of the *Sukambasi Ayog* were transferred to the Ministry of Land Reform.

In 1996 the Ministry of Land Reform announced different proposals to address the problems of the same 5,000 landless and indebted *kamaiya*. These aimed to provide *kamaiya* with proper education facilities, family planning advice, soft loans and income generation schemes. The income generation schemes which were proposed included training in a skill or trade, and advice on various forms of animal husbandry such as pig and rabbit farming. In addition there was a proposal to provide *kamaiya* with a piece of government land for a period of 15 years by which time they would in theory have become self sufficient. A controversial part of the project however was the proposal that significant expenditure should go to landlords in repayment of the *kamaiya's* debts. This led to criticism from NGOs and confusion as to who would be the real beneficiaries of the scheme.

A Minimum Wage

In 1992, the Government introduced the Labour Act, which replaced the previous Factory and Factory Workers' Act (1959). While the Labour Act makes some provision for minimum wages and their enforcement, this deals almost exclusively with the industrial sector. Therefore, there is, effectively, no minimum wage for the vast majority of Nepal's labour force.

The Ministry of Labour, in cooperation with the Centre for Economic and Technical Studies and the International Labour Organization (ILO), conducted a study on minimum wages for agricultural workers in Nepal from 1 September to 31 December 1995.

The study defined the minimum wage as being the minimum daily rate for subsistence including food, clothing and medicine. It attempted to set levels for a minimum wage in both the hills and *terai* on this basis. It suggested a level of NR 60 (US$1.08) in the hills and NR 50 (US$0.90) in the *terai*. The actual levels prevailing in these areas were, however, found to be substantially lower.

Land Reform

In theory parties of both the left and the right are committed to land reform. This is a popular subject with the voting public and has become a political battleground. Each party interprets the term 'land reform' differently, however, and all but the CPN UML stop short of advocating land redistribution.

The CPN UML Government, which came to power at the end of 1994, made land reform the centrepiece of its development policy and set up a commission to develop detailed proposals. The Land Reform Commission estimated there to be two million agricultural labourers in Nepal who worked on land other than their own, and identified this group as being in extreme poverty and a priority for government assistance. It proposed to distribute enough land to the families of landless agricultural labourers to allow them to sustain a household and also allow for a small surplus for sale. The rationale was that this would increase the buying power of a large section of the rural populaton and encourage the growth of a new market for agricultural products with added value, other than the basic staples. It was argued that this would help promote agro-business, creating alternative employment opportunities for former agricultural labourers and creating business opportunities in which former landlords would be willing to invest.

A large part of the land targeted for redistribution was to come from uncultivated areas owned by the Government. In addition, the maximum amount of land which any landlord was legally allowed to hold was to be reduced, and the surplus purchased with state funds for redistribution to the landless. The compulsory purchase of these lands by Government would force large landowners to liquidate some of their capital which they would be encouraged to invest in agro-industrial projects.

The proposals seemed to promise, at least in theory, some positive and fundamental changes which would directly address the problems of *haliya* and *kamaiya* bonded labourers. Whether or not the reforms would have been successful is academic. The Commission submitted its report shortly before the CPN UML lost power in 1995 and it was not published. At the time of writing, it remains to be seen whether the new coalition Government will revive the planned reforms. The Nepali Congress and RPP coalition Government formed in 1996 had less ambitious proposals for land reform, the main thrust of which was to give outright ownership of a proportion of the land currently under tenancy arrangement directly to the tenant farmer and do away with some of the exploitative relationships which had existed previously in the system of share-cropping. The Ministry of Land Reform did not intend to grant ownership or tenancy rights to landless agricultural labourers, and, therefore, the reforms would not have benefited *haliya* and *kamaiya* bonded labourers.

Immediately after the re-establishment of democracy popular expectations were high. The initial euphoria - some would say inevitably - has since given way to disillusionment. Certainly, the recent succession of minority and coalition governments has led to weak government and produced policies which, though they may be politically expedient, do not provide long-term solutions to bonded labour. Attempts to address bonded labour have been consequently frustrated by each side alternately seeking to diminish or undo the positive achievements of the other. The impasse created by the political system puts even more responsibility on NGOs to develop a national campaign which not only helps to direct government action against human rights violations along lines which will bring about solutions, but makes a solution to the problem a political prize worth winning. The recommendations set out in the concluding chapter have these specific goals in mind.

NOTES

1 Interview conducted by ASI/INSEC in September 1995.

2 Skar, H.O, 1991, p 32, *op.cit.*

3 Landless Settlers' Problem Resolution Commission, 1995, p 12, *op.cit.*

4 *ibid.*

Part IV
Conclusions and Recommendations

CHAPTER 11

Conclusions and Recommendations

Village society in Nepal is divided. This division has its basis in the caste system and in landholding structures. It even crosses over into the household unit itself where women traditionally have a servile status.

Policy makers and planners have tended to think of the Village Development Council (VDC) as being the basic unit of Nepal's development strategy without considering the effects of development on different groups of villagers. Consequently development initiatives have often failed to address the problems of the weakest and most marginalised members of the village who are unlikely to be represented in the decision-making process. Instead, development initiatives have tended to reinforce the power of the landed, high-caste elite at the local level.

Even when these divisions are recognised by planners, there is a tendency to believe that they will simply disappear with development. This is a dangerous form of complacency; 40 years of 'development' in Nepal have left the old structures more-or-less intact. They have proved resistant to change, even since the establishment of a multi-party political system. More significantly, rather than being considered a product of under-development, caste and landholding structures (and the discrimination and exploitation they entail) should in fact be regarded as a major cause. They have acted as a huge brake on the development of Nepal's human resources and the consequent growth of the rural economy. The very people charged with implementing development initiatives at the village level usually come from an elite which derives its power from keeping the poorest sections of society dependent and in poverty.

The issue of bonded labour is being addressed by some NGOs. Efforts so far, however, have focused only on the *Kamaiya* system in five districts of the western *terai*. The resulting programmes have been highly effective in empowering the *kamaiya* groups concerned, publicising the issue, and putting pressure on government for action. The research undertaken by INSEC and ASI has established that systems of debt bondage and serfdom are not limited to one remote area of the country, but exist in many different districts in slightly different forms.

On this scale, bonded labour undermines the basic freedoms and rights of a significant proportion of the population, frustrating the development of a multi-party democracy. There is a clear obligation on the Nepali Government to take action to end this gross abuse of human rights and a corresponding obligation on the wider international community to support such moves. The experience in both India and Pakistan has shown the importance of introducing national legislation to abolish bonded labour and other slavery-like practices. While in both cases the government authorities concerned have often been ineffectual and selective in their implementation of such legislation, the law has been used to great effect by NGOs. Furthermore, it provides an essential framework for channelling public resources into the problem and gives direction to the relevant government agencies.

For these reasons the main recommendations of this report revolve around the introduction of domestic legislation abolishing bonded labour and other slavery-like practices.

I. Recommendations to the Government of Nepal

A Abolish Bonded Labour and other Slavery-like Practices.

In 1963 Nepal ratified the Slavery Convention (1926) and the Supplementary Convention on the Abolition of Slavery, the Slave Trade, and Institutions and Practices Similar to Slavery (1956).

The latter specifically prohibits the practices of bonded labour and serfdom. The Government is under an obligation to the international community to take action to implement these international human rights treaties, but has so far not prohibited bonded labour by law. Consequently, it is failing to meet its international commitments.

A.1. The Government of Nepal should show its commitment to the abolition of both debt bondage and other forms of forced labour by ratifying the International Labour Organization's Convention No. 29 concerning Forced Labour (1930), which has been ratified by 135 countries. This is one of the ILO's 'core conventions', and failure to ratify it implies a lack of commitment to upholding some of the most basic labour standards.

A.2. The Government of Nepal should introduce a law banning all slavery-like practices as an essential first step towards eradication of the problem. This law should ban all institutions and

practices akin to slavery according to the definitions contained in the UN Supplementary Convention on Slavery (1956).

It should make provision for the identification of the victims, for their rehabilitation and for the punishment of landlords or other employers who exploit bonded labour in the future. It should also have the effect of cancelling any debts which currently bind individuals or families to work for their creditor.

The Government should act swiftly to expedite hearing of the case which has been taken to the Supreme Court challenging the constitutional commitment to prohibit and punish slavery and slavery-like practices.

B. Land Reform

While legislation against slavery-like practices is an essential first step, by itself it will not be enough to bring an end to bonded labour. The victims of bonded labour and serfdom must be allowed the means to generate an income of their own which would be sufficient to sustain a family and to avoid the need to take on further debts and slip back into slavery.

B.1. The Government of Nepal should make land or other resources available to provide for the rehabilitation of the victims of slavery, enabling them to use their existing skills and knowledge to support themselves and their families. Areas of unused government land suitable for this purpose have already been identified by the Ministry of Land Reform. The compulsory purchase of lands currently held in excess of the legally permitted maximum would generate an additional area of land for redistribution.

C. Minimum Wage

While the Labour Act of 1992 provides for minimum wages and conditions of work in the industrial sector, the Act does not extend to agriculture, except in the case of some government-owned plantations. According to the *Muluki Ain* (Civil Code) of 1991, conditions of work and rates of pay are a matter for the employer and employee to decide, even when an employee is working for no payment. No legal minimum wage is applicable to agricultural labour.

C.1. The Government of Nepal should introduce a minimum wage for agricultural labour, set at no less than the minimum daily rate for subsistence including provision for food, clothing and medical expenses.

D. Caste Discrimination

Discrimination on the grounds of caste is illegal in Nepal. However, the law is routinely ignored in this respect. People from 'untouchable' castes, especially those from rural areas, are routinely excluded from playing an equal role in local decision-making processes and are marginalised from society. This leaves these groups vulnerable to slavery-like practices. In addition 'untouchable' groups are prevented by custom from following occupations outside their traditional caste roles and are forced instead to perform servile tasks, sometimes without payment.

D.1. The Government of Nepal should act to ensure that the laws regarding caste discrimination are properly enforced in rural areas. This requires the adoption of new procedures or the establishment of a special commission to promote equality between castes and ethnic groups in Nepal.

E. Credit for Medical Costs

From the interviews and group discussions conducted in the course of this research it became clear that a major reason why bonded labourers originally resorted to informal sources of credit was to pay for the treatment of tuberculosis and other serious medical conditions. Such unregulated sources of credit generally incur very heavy rates of interest and can lead to debt bondage.

E.1. The Government of Nepal should introduce measures to help the rural poor find alternative ways of paying for medical facilities, such as a system of deferred payment or regulated credit.

II. Recommendations to Nepalese Non-Governmental Organisations (NGOs) and Trade Unions, and International Non-Governmental Organisations (INGOs) Working in Nepal

A. Advocacy

A.1. NGOs and trade unions should campaign to persuade the Government to take action against bonded labour. There is already a strong national level grouping of NGOs and trade unions working to focus government attention on bonded labour. This is an ideal platform from which to make concerted appeals to the Government of Nepal to introduce legislation abolishing slavery-like practices in all areas of the country. Joint actions could be undertaken to promote debate

about bonded labour of haliya agricultural labourers, and to coordinate further research as required.

In addition, appeals should be made to the Government of Nepal to introduce positive discrimination in recruitment of 'untouchable' and low-status ethnic groups in education, government employment and law enforcement. Campaigning against the practice of 'untouchability' should be intensified, especially in rural areas, and linked directly to the problems of bonded labour.

B. NGO Programmes

B.1. NGOs have built up a great deal of experience in addressing the *Kamaiya* system of bonded labour in Kanchanpur, Kailali, Bardiya, Banke and Dang. NGOs should consider expanding these programmes, initially to cover one or two districts in the hills and *terai* where other forms of bonded labour are prevalent. On the basis of what has proved successful so far, these programmes should include:

a) Awareness raising, group formation and human rights education, through the introduction of non-formal literacy programmes and evening classes.

b) Income generation, credit and savings schemes.

c) Special attention to the particular situation of women in bonded labour families.

III. Recommendations to Governments and Others Outside Nepal

A.1 Governments and inter-governmental agencies should urge the Government of Nepal to introduce domestic legislation explicitly prohibiting bonded labour and any other practices akin to slavery, and thus honour the international commitments it has undertaken in this respect.

B.1. Governmental and inter-governmental development agencies operating in Nepal should offer technical assistance to the Government of Nepal in the drafting and implementation of such legislation. They should work together with relevant ministries and relevant local non-governmental organizations to develop programmes to rehabilitate the groups concerned.

C.1. **Trade union organisations in particular should urge the Government of Nepal to ratify the International Labour Organization Convention No. 29 concerning forced labour. They should also support appeals to the Government of Nepal to introduce a minimum wage for agricultural labour. They should support local trade unions such as GEFONT and the Nepal Confederation of Trade Unions in programmes to organise agricultural workers in Nepal in their campaign against all forms of slavery.**

Appendix 1
Weights, Measures and Currency

Area

1 *Bigha*	=0.67 hectare	=1.7 acres
1 *Ropani*	=0.034 hectare	=0.085 acre
1 *Nali*	=0.01 hectare	=0.03 acre
1 *Katta*	=0.034 hectare	=0.085 acre

Weights (*approximate measures*)

1 *Quintal*	=100 kg
1 *Mon*	=40 kg
1 *Bora*	=sack weighing between 75-100 kg
1 *Mana*	=0.5 kg
1 *Pathi*	=4 kg (husked rice)

Currency – Exchange Rates

Where currency appears in the text equivalents in US$ have been shown accurate to the month when the research was carried out. For this reason the following exchange rates may not correspond exactly to the equivalents stated in the text.

Nepali R 100 = Indian R 60 = US$ 1.80 = £stg 1.20

Appendix 2
Glossary

Afno Manche - 'one's own people', a system of extended nepotism where decisions over jobs, contracts, or political favours are decided on the basis of family or caste connections, rather than objective criteria based on the ability or qualifications of the persons concerned.

bigha - Nepali system of land measurement equal to 0.67 of a hectare of land. Traditionally a *kamaiya* would receive the produce of one *bigha* from the landowner as payment for his work during the agricultural year. The term has now come to mean any in-kind payment made to a *kamaiya*.

birta - from the Nepali word meaning 'hero'. Large areas of land given by kings and Rana rulers to successful generals or nobles in order to secure their loyalty.

Brahmin - priestly caste. The highest of the four main divisions, or *varna* (see below) of the Hindu caste system.

Brahminism - the formalised practice of orthodox Hindu religion.

bukhra - the name given to the hut where *kamaiya* live on landlord's land.

bukrahi - female *kamaiya*.

Chamar - a caste found mainly in the *terai*, associated with very low-status work such as latrine cleaning and designated 'untouchable' in the Nepali caste system.

Chaudhary - originally applied to headmen of the Tharu ethnic group, now taken to mean any Tharu person, and frequently used as a last name by Tharu people.

Chepang - ethnic group from areas in the hills and the *terai*. Considered to be of low status in the Nepali caste system.

Chhetri - traditionally the warrior caste, also known as *kshatriya*. The second of the four major divisions, or *varna*, of the Hindu caste system, after *Brahmins*.

CPN UML - Communist Party of Nepal, United Marxists and Leninists.

dalit - 'the oppressed' widely used in India in preference to the pejorative term 'untouchable'. Some groups are now using this term in Nepal.

Damai - a caste of tailors, found throughout the country, designated 'untouchable' in the Nepali caste system.

Dhangar - an ethnic group from the *terai*, designated 'untouchable' in the Nepali caste system.

Dhimal - see Dhangar

Dushad - a caste found mainly in the *terai*. Associated with low - status work such as funeral preparations and disposal of carcasses, and designated 'untouchable' in the Nepali caste system.

Dom or **Doom** - a name used by

some high-caste people as a general insult for all low-caste 'untouchable' groups. It is also used to refer to the settlements of 'untouchable' communities.

haliya - literally 'ploughman', used to refer to male agricultural labourers. It is associated with exploitative practices such as debt bondage and serfdom in Nepal and India.

Halkhor - a caste found mainly in the *terai*, associated with very low-status work such as latrine cleaning and designated 'untouchable' in the Nepali caste system.

hari - a term synonymous with *haliya* (see above) used in certain parts of Nepal and also in Pakistan.

haruwa - a term synonymous with *haliya* (see above) used in certain parts of Nepal and also used in India.

jagir - literally 'official employment'. It referred in particular to a system of payment to public servants where instead of a salary, they received a grant of land. The system was abolished as part of the land reforms of the 1950s and 1960s.

jamindar (also known as **zamindar**) - originally a term referring to the intermediaries of landlords, empowered to deal with all aspects of land administration. Nowadays the term is applied to any large landowners.

jutakochala - literally "garland of shoes" - a rite of public humiliation. In Nepal shoes are a powerful symbol of all that is low and dirty. If a member of a village is seen to have cheated his neighbours, let them down or disgraced him or herself in their eyes, a garland, made from old shoes is placed over their head. This is seen as the ultimate disgrace and often results in that person having to leave the village.

kamaiya - a system of bonded labour that exists mainly in five districts in western Nepal. The term is also applied to the bonded labourers themselves.

Kamaiya Mukti Manch - the Kamaiya Liberation Forum. An organisation of *kamaiya* and *kamaiya* activists dedicated to the abolition of the *Kamaiya* system.

Kami - a caste of blacksmiths found throughout Nepal but sometimes known by other names (such as *Lohar* in the west). Designated 'untouchable' in the Nepali caste system.

khas - peoples who migrated into the western hill regions of Nepal from the Indo-Gangetic plain. Credited with introducing the current Nepali language. People now known as *Pahariya* (hill people) are assumed to have *Khas* ancestry.

Khataves - a caste from the eastern *terai* associated with ditch digging and agricultural labour. Designated 'untouchable' in the Nepali caste system. This caste name is often applied to the *Musahar* ethnic group (see below).

kisan - a farmer of a small holding.

Kshatriya - see **Chhetri**.

Lohar - see **Kami**.

Maithili - a dialect spoken in the *terai*, very similar to Hindi.

Magh - month covering the period from 14 January to 11 February.

Maghi - festival held on the first day of *Magh*.

masyaura - a payment in kind given by landlords to *kamaiya*.

Muluki Ain - civil code laid down originally by Jung Bahadur Rana in 1854.

Musahar - a tribal group from the eastern *terai*, thought to have migrated originally from the Indian states of Bihar and West Bengal. Associated with agricultural work such as ditch digging for the construction and maintenance of rice fields. Designated 'untouchable' in the Nepali caste system (see also *Khataves* above).

Newar - an ethnic group of Tibeto-Burman ancestry located mainly in the Kathmandu Valley. The Newars have their own language and their own caste system and are associated with the very early development of Hindu religion in the Kathmandu Valley.

NFE - non-formal education. Literacy classes which take place outside of schools to allow adults and working children flexible access to education.

NGO - Non-governmental organisation.

orgoniya - young unmarried girls who work as bonded domestic servants in their landlord's house under the *kamaiya* system.

Pahariya - 'hill people'. The term refers to Nepali-speaking people descended from the Khas (see above).

panchayat - a system of local councils, introduced as a partyless system of government in 1962.

peon - office messenger.

rakam - a system of payment to state employees similar to the *Jagir* system (see above).

Sarki - a caste of leather workers found throughout Nepal and given 'untouchable' status in the Nepali caste system.

Satar - tribal people located in the eastern *terai* and also found in India. Designated as 'untouchable' in the Nepali caste system.

saunki - the debt that a *kamaiya* takes on.

Shudra - the lowest of the four main divisions or *varna* in the orthodox Hindu caste system consisting of occupational castes and trades.

Sukambasi Ayog - Landless Settlers' Problem Resolution Committee. A committee set up by HMG Nepal on the question of landless people.

Sunar - an occupational caste of goldsmiths, designated as 'untouchable' in the Hindu caste system.

Tagadhari - 'wearers of the Holy Cord', signifying a high-caste status.

Tatmas - caste of 'untouchable' status from the *terai*, associated with agricultural work.

terai - narrow lowland strip bordering the Indo-Gangetic plain. The most fertile region of Nepal.

Tharu - ethnic group from the western *terai*. Considered to be of low 'enslavable' status in the Nepali caste system but not 'untouchable'. The vast majority of bonded labourers involved in the *Kamaiya* system of bonded labour are from the Tharu ethnic group.

Vaiysha - third of the four main divisions (or *varna*) of the Hindu caste system. Traditionally businessmen and farmers.

varna - in the orthodox Hindu caste system the population is divided into four main groups or *varna*, *Brahmin*, *Kshatriya*, *Vaiysha*, and *Shudra*. These broad divisions are then subdivided into a range of different castes.

VDC - Village Development Committee. The main local administrative unit in Nepal. Its boundaries include the village and its outlying settlements. This area is administered by a committee made up of elected members presided over by the VDC chairman, a powerful local figure.

Appendix 3
Select Bibliography

Agricultural Projects Services Centre & John Melrose Associates Inc, 1995. *Nepal: Agriculture Perspective Plan - Final Report*, (prepared for National Planning Commission His Majesty's Government of Nepal & Asian Development Bank), Kathmandu.

Backward Society Education (BASE), 1994. *Tharu Education for Transformation, Third Year Report*, (submitted to Danida), Dang.

Bista, D.B, 1991. *Fatalism and Development: Nepal's Struggle for Modernization*, Orient Longman Limited, Calcutta.

Bista, D.B, 1967. *People of Nepal*, Ratna Pustak Bhandar, Kathmandu.

Central Bureau of Statistics, 1993. *National Sample Census of Agriculture, Nepal 1991/92*, His Majesty's Government National Planning Commission Secretariat, Kathmandu.

Central Bureau of Statistics, 1994. *National Sample Census of Agriculture, Nepal 1991/92 - Analysis of Results*, His Majesty's Government National Planning Commission Secretariat, Kathmandu.

Central Bureau of Statistics, 1995. *Population Monograph of Nepal*, His Majesty's Government National Planning Commission Secretariat, Kathmandu.

Central Bureau of Statistics, 1995. *Statistical Year Book of Nepal 1995*, His Majesty's Government National Planning Commission Secretariat, Kathmandu.

Cox, T, 1994. *The Current Socio-economic Status of Untouchables in Nepal*, Occasional Papers in Sociology and Anthropology, Vol.4 , Tribhuran University, Kathmandu.

Gautam, R, and Thapa -Magar, A.K, 1994. *Tribal Ethnography of Nepal* Vols I and II, Book Faith India, Delhi.

Gurung, G.M, 1994. *Indigenous Peoples: Mobilisation & Change*, HISI Press, Kathmandu.

Hofer, A, 1979. *The Caste Hierarchy and State in Nepal: A Study of the Muluki Ain of 1854*, Universitätsverlag Wagner, Transbrück.

INSEC, 1992. *Bonded Labour in Nepal Under* Kamaiya *System*, INSEC, Kathmandu.

INSEC, 1994. *Human Rights Year Book 1993*, INSEC, Kathmandu.

Landless Settlers' Problem Resolution Commission, 1995. *Survey on Landless Bonded Labour (Kamaiya): A Report*, His Majesty's Government of Nepal, Kathmandu.

Mishra, S, 1994. *Voice of the Voiceless* (video), INSEC, Kathmandu

NCIY, 1993. *Indigenous Peoples of Nepal*, (Papers prepared for the UN's World Conference on Human Rights in Vienna, Austria, 14 -25 June 1993), Kathmandu.

Nepal Rastra Bank, 1994, *Nepal Rural Credit Review: Final Report*, Nepal Rastra Bank, Kathmandu.

Oxfam, 1990. *Nepal Programme Review*, Oxford.

Ram Bahadur, K. C, 1986. *Land Reform in Nepal, His Majesty's Government of Nepal*, USAID, GTZ & Winrock.

Rankin, K, forthcoming in Skar, H.O (Ed.), *Tharu and Tarain Neighbours*, in press.

Sattaur, O, 1993. *Child Labour in Nepal*, Anti-Slavery International and CWIN, Kathmandu.

Skar, H.O, 1991. *Nepal, Indigenous Issues and Civil Rights: The Plight of the Rana Tharu*, Norwegian Institute of International Affairs, NUPI Report, Oslo.

Sustainable Livelihood Forum, 1993. *Annual Report I and II 1991–1992*, Kathmandu.

Appendix 4
NGOs in Nepal Working on Bonded Labour

Backward Society Education (BASE)
Project Office Tulsipur Dang
Tel: (00 977) 82 20043
Fax: (00 977) 82 20312

Child Workers in Nepal - Concerned Centre (CWIN)
PO Box 4374, Kathmandu
Tel: (00 977) 1 270336
Fax: (00 977) 1 224466

General Federation of Nepalese Trade Unions (GEFONT)
PO Box 10652, Kathmandu
Tel: (00 977) 1 411880
Fax: (00 977) 1 420547
e-mail: gefont@wlink.com.np

GRINSO
PO Box 5670, Kathmandu
Tel: (00 977) 1 420904
Fax: (00 977) 1 424405

Informal Sector Service Centre (INSEC)
PO Box 2726, Kathmandu
Tel: (00 977) 1 270770/272267
Fax: (00 977) 1 226820
e-mail: kalanki@insec.wlink.com.np

Kamaiya Liberation Front - Nepal
PO Box 10652, Kathmandu
Contact Tel: (00 977) 1 411880
Fax: (00 977) 1 420547

Nepal Federation of Nationalities (NEFEN)
PO Box 822, Kathmandu

RRN
PO Box 8130, Kathmandu
Tel: (00 977) 1 415418
Fax: (00 977) 1 418296

SAP Nepal
PO Box 3827, Kathmandu
Tel: (00 977) 1 476163
Fax: (00 977) 1 471779

Society for Participatory Cultural Education (SPACE)
PO Box 1668, Kathmandu

Selected publications from Anti-Slavery International

Ethnic Groups in Burma:
Development, Democracy and Human Rights
by Martin Smith (1994) £5.95

Slavery in Brazil: A Link in the Chain of Modernisation
by Alison Sutton (1994) £5.50

Britain's Secret Slaves:
The Plight of Overseas Domestic Workers in the United Kingdom
by Bridget Anderson (1993) £5.50

This Menace of Bonded Labour:
Debt Bondage in Pakistan (1996) £2.99

Enslaved Peoples in the 1990s:
Indigenous Peoples, Debt Bondage and Human Rights (1997) £5.95

Postage and packing: add 50p for each item sent to address in the UK, and £1 worldwide (airmail).

Selected publications from INSEC

Informal (regular tri-monthly bulletin)

Human Rights Situationer (quarterly newsletter)

Human Rights Year Book (1996)

(Prices and cost of postage available on request)

Join
Anti-Slavery International
and support our campaign against bonded labour and all other contemporary forms of slavery.

Keep up to date with news and action on the issues with our quarterly journal, *The Reporter* – free to all members.

Individuals: £25 a year (UK and overseas)
Students/unwaged: £8 (UK only)
Donations welcome

name ______________________________

address ______________________________

country/post code ______________________________

Please enclose a cheque or money order payable to Anti-Slavery International, or give your credit card details here:
Access/ Visa/ Mastercard/ Amex/ Diners' Club/ CAF

number ______________________________

expiry date ______________________________

amount paid ______________________________

signature ______________________________

date ______________________________

☐ Please send me the current issue of *The Reporter*

☐ I would like a complete list of ASI publications

Return to:

Anti-Slavery International,

The Stableyard, Broomgrove Road,

London SW9 9TL, United Kingdom.

Registered charity 1049160

Informal Sector Service Centre
Informal Sector Service Centre (INSEC) is an independent, non-profitmaking and non-partisan human rights organisation emphasising social justice. Based in Kathmandu and in all the regional centres of Nepal, the organisation was founded in 1988 to provide services to grassroots organisations or individuals actively involved in promoting human rights.

INSEC is committed to the protection and promotion of human rights through advocacy, lobbying, training, education and information campaigns at national, regional and grassroots levels. Specific activities include:

- human rights monitoring through district representatives and periodic fact-finding missions published in the Human Rights Year Book and other publications;
- human rights education through programmes in 34 districts.
- combatting all forms of slavery with a special focus on *kamaiya*;
- promotion of women's rights;
- intervention in human rights law-making.

Anti-Slavery International
Anti-Slavery International (ASI) promotes the eradication of slavery and slavery-like practices, and freedom for everyone who is subjected to them. The abuses which ASI opposes include: slavery and the buying and selling of people as objects; trafficking of women and the predicament of migrant workers who are trapped into servitude; debt bondage and other traditions which force people into low status work; forced labour; forced prostitution; abusive forms of child labour; and early or forced marriage and other forms of servile marriage. ASI focuses on the rights of people who are particularly vulnerable to exploitation of their labour, notably women, children, migrant workers and indigenous peoples.

ASI pursues its objectives by:

- collecting information about these abuses, bringing them to the attention of the public and
- promoting public action to end them; identifying ways in which these abuses can be brought to an end, and influencing policy-makers in governments or other institutions at national and international level to take action
- accordingly; supporting victims of the abuses which ASI opposes in their struggle for freedom, in particular by working with organisations they establish and other organisations campaigning on their behalf.